Using Social Media to Extract Information About Chemical Weapons Incidents

A Methodology and Demonstration of Concept from the Civil War in Syria

JOSHUA MENDELSOHN, STEPHANIE YOUNG,
JENNY OBERHOLTZER, YOUSUF ABDELFATAH,
GREGORY WEIDER FAUERBACH, NATASHA LANDER,
PAUL S. STEINBERG

Prepared for the Defense Threat Reduction Agency
Approved for public release; distribution unlimited

For more information on this publication, visit www.rand.org/t/RRA620-1

Library of Congress Cataloging-in-Publication Data is available for this publication.
ISBN: 978-1-9774-0644-6

Published by the RAND Corporation, Santa Monica, Calif.

Cover design by Rick Penn-Kraus;
city image: petrovv/Getty Images/iStockphoto;
icons: soulcld/Getty Images

Preface

Policymakers across the federal government have begun to recognize the potential of social media as a source of information and have commissioned studies to explore how social media can improve disaster situational awareness, influence public opinion, augment traditional data sources, and counter disinformation. In this project, we developed an approach for analyzing social media data to derive insights about chemical incidents and conducted a proof of concept of that approach by applying it to the case of chemical weapons employment in Syria between 2017 and 2018.

This report's findings will be of interest to policymakers considering avenues for deriving intelligence from social media and to data scientists tasked with building this capability.

This research was sponsored by the Defense Threat Reduction Agency (DTRA) and conducted within the International Security and Defense Policy Center of the RAND National Security Research Division (NSRD), which operates the National Defense Research Institute (NDRI), a federally funded research and development center sponsored by the Office of the Secretary of Defense, the Joint Staff, the Unified Combatant Commands, the Navy, the Marine Corps, the defense agencies, and the defense intelligence enterprise. The research reported here was completed in September 2020 and underwent security review with the sponsor and the Defense Office of Prepublication and Security Review before public release.

For more information on the RAND International Security and Defense Policy Center, see www.rand.org/nsrd/isdp or contact the director (contact information is provided on the webpage).

Contents

Figures

Tables

Summary

In this report, we describe a procedure for analyzing chemical weapons incidents and apply that method to the test case of chemical weapons use in the Syrian Civil War.

Our procedure is a four-step process:

- Step 1. Identify operationally relevant factors and examine known events to find incident indicators. To support this step in our test exercise, we consulted with subject-matter experts and reviewed YouTube, Twitter, and Facebook posts from 19 chemical weapons incidents in Syria. We developed keyword lists for Step 3 and analyst checklists for Step 4 (described below).
- Step 2. Develop a feed of social media data. To support this step in our test exercise, we generated a statistically representative sample of 206,423 Twitter users who tweet in Arabic and downloaded all the tweets they wrote in the 2017–2018 period.
- Step 3. Conduct automated daily scans for elevated keyword use in Twitter data. To support this step, we divided the keywords from Step 1 into five groups—angry reactions, lamentations, chemical agent names, attack descriptors, and exposure symptoms—and searched for elevated keyword usage for any of the groups on any of the 730 days in our data set.
- Step 4. Analyze posts to verify detection and extract information. To support this step, we built a playbook with six techniques for extracting information from social media posts—multiple post consistency check, exact phrase search, image search, geographic consistency check, source tracing, and links and mentions tracing.

In our Syrian Civil War test case, the procedure showed promise:

- Our elevated keyword use detection algorithm achieved recall rates of 40 percent at predicting when actual events occurred—not good enough for operational use but promising for a procedure in its developmental infancy.
- In a double-blind trial, our tester analyst found that tweets with subject-matter keywords, written on a day with elevated keyword usage, had relevant information 84 percent of the time.
- Using the posts, the analyst was consistently able to identify information related to the who, what, and where questions about the chemical weapons incident.

We recommend that the Defense Threat Reduction Agency initiate three activities to further the development of this procedure.

- Develop resource-pooling agreements with other federal agencies. The procedure may be more likely to achieve cost-effectiveness as a shared multipurpose resource.
- Build an extensive chemical weapons keyword list and rigorously test it against best available intelligence. The elevated keyword detection algorithm may be more likely to achieve high recall rates if developed further in this way.
- Conduct an exercise with federal computers and staff, scanning for easily verified types of events. This exercise may strengthen the evidence base on the procedure's effectiveness. It will also be a chance to write procedural documents and muster other resources needed to make this capability operational.

Decisionmakers should consider three things in planning for operational implementation.

1. Is the local social media environment suitable? This procedure is most appropriate in places with many active social media users and little state-supported censorship or disinformation. Decisionmakers should also consider the social media platform's terms of use and the cultural competency of available analysts.

2. Are enough staffing resources available? We estimate that it may require 70 to 230 full-time analysts to conduct timely analysis of posts in one language on days when elevated keyword use is detected. However, if these analysts were a shared resource—only incurring staffing costs on days when elevated keyword use is detected—the actual annual staffing costs might only be the equivalent of one to 29 full-time analysts.
3. Are support resources in place? Repeated exposure to chemical weapons incidents on social media can lead to secondary traumatic stress. This report includes appendixes of detailed strategies for mitigating analyst traumatic stress risk.

Acknowledgments

We acknowledge the support of the sponsor, the Defense Threat Reduction Agency Strategic Trends and Effects Department, for this work. In particular, we thank Robert Peters, Donald Wenzlick, and Jennifer Perry, as well as Christine Wormuth and Michael Spirtas of the International Security and Defense Policy Center within the RAND National Defense Research Institute for their support. We also acknowledge RAND colleagues who contributed to data collection, quality assurance review, and development of material related to secondary trauma, including David Kravitz, Lisa Jaycox, Terri Tanielian, Kirsten Becker, Lynsay Ayer, Dionne Barnes-Proby, Coreen Farris, Liz D'Amico, and Joan Tucker. Finally, we thank Elizabeth Bodine-Baron of RAND and Sean Everton of the Naval Postgraduate School for their thoughtful reviews. Any errors remain the responsibility of the authors alone.

Abbreviations

API	application programming interface
CBRN	chemical, biological, radiological, and nuclear
DSM-5	*Diagnostic and Statistical Manual of Mental Disorders, 5th Ed.*
DTRA	Defense Threat Reduction Agency
FN	false negative
FP	false positive
NDS	National Defense Strategy
NDRI	RAND National Defense Research Institute
OPCW	Organization for the Prohibition of Chemical Weapons
SME	subject-matter expert
STE	staff time equivalent
TOS	terms of service
TN	true negative
TP	true positive

CHAPTER ONE

Introduction

Background

Policymakers across the federal government are recognizing the potential of social media as a source of information and have commissioned studies to explore how it can improve disaster situational awareness, influence public opinion, augment traditional data sources, and counter disinformation.[1] Social media can be collected as soon as it is generated, can tap into a massive pool of potential observers, and is remotely accessible. However, because social media reflects the idiosyncratic interests and circumstances of each platform's user base, it lacks the rigor or systematic focus of professional reporting.[2] Given this

---

[1] See, for example, Cloé Gendronneau, Arkadiusz Wisniowski, Dilek Yildiz, Emilio Zagheni, Lee Florio, Yuan Hsiao, Martin Stepanek, Ingmar Weber, Guy Abel, and Stijn Hoorens, *Measuring Labour Mobility and Migration Using Big Data: Exploring the Potential of Social-Media Data for Measuring EU Mobility Flows and Stocks of EU Movers*, Brussels: European Commission, 2019; Todd C. Helmus, Elizabeth Bodine-Baron, Andrew Radin, Madeline Magnuson, Joshua Mendelsohn, William Marcellino, Andriy Bega, and Zev Winkelman, *Russian Social Media Influence: Understanding Russian Propaganda in Eastern Europe*. Santa Monica, Calif.: RAND Corporation, 2018; Jennie W. Wenger, Heather Krull, Elizabeth Bodine-Baron, Eric V. Larson, Joshua Mendelsohn, Tepring Piquado, and Christine Anne Vaughan, *Social Media and the Army: Implications for Outreach and Recruiting*, Santa Monica, Calif.: RAND Corporation, RR-2686-A, 2019; Douglas Yeung, Sarah A. Nowak, Sohaela Amiri, Aaron C. Davenport, Emily Hoch, Kelly Klima, and Colleen M. McCullough, *U.S. Coast Guard Emergency Response and Disaster Operations: Using Social Media for Situational Awareness*, Santa Monica, Calif.: RAND Corporation, RR-4296-DHS, 2020.

[2] Sue E. Kase, Elizabeth K. Bowman, Tanvir Al Amin, and Tarek Abdelzaher, *Exploiting Social Media for Army Operations: Syrian Civil War Use Case, Proceedings of SPIE (the*

limitation, research is needed to determine the circumstances under which social media can be an effective information source for different applications and what procedures are effective for distilling that information. To explore one such potential application, the Defense Threat Reduction Agency (DTRA) asked the RAND National Defense Research Institute (NDRI) to develop a method for effectively analyzing chemical threat incidents and to apply that method to the test case of chemical weapons use in the Syrian Civil War.

By the time civil war broke out in 2011, Syria's chemical weapons program was reported to be one of the largest and most advanced in the region, and much international attention focused on monitoring and securing the country's chemical weapons. Articulating related concerns, President Barack Obama famously drew a red line around the use or proliferation of chemical weapons in July 2012.[3] Despite President Obama's red line, reports of chemical weapons use began to emerge in December 2012 in an attack identified as the regime's first—against the city of Homs.[4] Another reported attack occurred on March 19, 2013.[5] When a sarin gas attack struck Ghouta in August 2013, killing an estimated 1,400 people, firsthand reporting posted to social media platforms was some of the earliest eyes on the ground.[6] At approximately 2:30 a.m., the first reports began to appear on social media, and, in the hours that followed, thousands of similar reports emerged from at least 12 different locations. By dawn, media outlets around the world had picked up the story, which was supported by

International Society for Optical Engineering), Vol. 9122, July 2014, p. 2; William Marcellino, Meagan L. Smith, Christopher Paul, and Lauren Skrabala, *Monitoring Social Media: Lessons for Future Department of Defense Social Media Analysis in Support of Information Operations*, Santa Monica, Calif.: RAND Corporation, RR-1742-OSD, 2017, p. 19.

[3] Barack Obama, "Remarks by the President to the White House Press Corps," webpage, August 20, 2012.

[4] Charles R. Lister, *The Syrian Jihad: al-Qaeda, the Islamic State and the Evolution of an Insurgency*, New York, Oxford University Press, 2016, p. 93.

[5] Lister, 2016, p. 153.

[6] At the time of the major sarin attack in August 2013 in Ghouta, a suburb of Damascus, a United Nations team was already on the ground investigating the March attack.

shocking images of the situation on the ground.[7] Despite the Syrian government's agreement in the aftermath of Ghouta to destroy its stockpiles, chemical weapons use did not end there. The use of chemical agents has also not been the sole purview of the Assad regime. The United Nations team investigating reported attacks since April 2014 attributed at least one to the Islamic State.[8] In total, some sources report more than 300 chemical weapons attacks.[9]

In addition to Syria being one of the world's most active hotspots for chemical weapons use throughout the 2010s, we use also the Syrian Civil War as a test case; the conflict is particularly salient because of the complex human terrain. In the Syrian Civil War, the United States has encountered threats from nonstate proxies and from states that enabled or failed to prevent chemical weapons proliferation.[10] DTRA director Vayl S. Oxford has emphasized that competing successfully against this range of activities requires defeating *threat networks* that facilitate "malign activities of state adversaries and their agents and proxies."[11] Features of social media data suggest that social media could be a valuable tool to counter threat networks. For example, social media's user base could provide a large, low-latency stream of information about ground conditions and human terrain. Such analysis could illuminate threat networks, thus providing insight into relationships and key nodes that enable operations.

---

[7] Brian Lessenberry, "Intelligence Integration and the Syrian CW Threat," Center for Strategic and International Studies webpage, February 18, 2015.

[8] Carla E. Humud, Christopher M. Blanchard, and Mary Beth D. Nikitin, *Armed Conflict in Syria: Overview and U.S. Response*, Washington, D.C.: Congressional Research Service, RL33487, January 6, 2017.

[9] Tobias Schneider and Theresa Lütkefend, *Nowhere to Hide: The Logic of Chemical Weapons Use in Syria*, Berlin: Global Public Policy Institute, February 2019.

[10] Vayl S. Oxford, "Countering Threat Networks to Deter, Compete, and Competition Below Armed Conflict with Revisionist Powers," *Joint Forces Quarterly*, No. 95, 4th Quarter 2019.

[11] Oxford, 2019, p. 79.

Objective of This Report and Research Approach

To determine the utility of social media analysis for this problem set, DTRA asked RAND NDRI to create a method for effectively analyzing chemical threat incidents and apply that method as a proof of concept to the use of chemical weapons in the Syrian Civil War. Four key research questions drove this analysis:

1. When a chemical event happens, what information about it tends to appear in social media, and what forms does that information typically take?
2. How can a computational approach enable rapid detection of chemical weapons incidents buried among millions of social media posts?
3. How can a blended computer-human approach improve on a fully automated approach?
4. What do leaders need to know to implement a social media analysis capability?

The methodology developed in our research contributes to the field of emerging event analysis—multidisciplinary efforts to use digital data like social media to detect, verify, and extract information about events more rapidly than traditional sources. See Appendix E for an overview of the applications and techniques of social media emerging event analysis.

We find that practitioners are exploring social media emerging event analysis in at least five topic areas. First, *marketing practitioners* strive to channel social media activity in ways that promote brands and link online activity to desired offline outcomes. Second, *public health practitioners* strive to educate the public and detect emerging health issues so that remedial action may be taken. Third, *disaster recovery practitioners* strive to establish situational awareness under circumstances where the flow of traditional information has been disrupted. Fourth, *public order and political influence practitioners* strive to gain situational awareness on emerging public safety incidents and social media activities that may lead to public safety incidents.

They may also seek to encourage pro-government viewpoints, shape public opinion, and suppress dissent, both domestically and abroad.[12] Finally, *misinformation researchers* strive to understand the spread of misinformation and its consequences for perceptions or behavior.[13]

Across the different topic areas, researchers typically apply one or more of the following data analysis techniques. First, *place name–surge techniques* search for abrupt increases in the use of proper names for a specific geographic location. Second, *activity-surge techniques* search for abrupt increases in the use of words that describe a specific kind of event. Third, *deductive machine-learning techniques* provide an algorithm with human-categorized examples of an event and then have the algorithm search for new events that it sees as similar to the examples of a specific category of events. Fourth, *inductive machine-learning techniques* infer categories by clustering posts according to their similarities and then determining the noteworthiness of emerging events based on how the new events fit within the categorization system. Fifth, *social-verification techniques* use interactions between users to determine the credibility of social media posts. Lastly, *blended computer-human intelligence technique*s apply computational filtering methods to identify a manageable subset of posts that warrant manual examination.[14]

Using input from DTRA subject-matter experts (SMEs), we selected a blend of techniques suitable for detecting a chemical incident and then applied those techniques to the case of chemical weapons employment in Syria in the 2017–2018 period. Our method follows a blended computer-human intelligence strategy, with manual keyword selection, automated keyword surge scanning to filter posts, and manual deep analysis of the filtered posts. Our keyword selection

---

[12] The overlap between public order activities and political influence activities can be very specific to national context. For governments that expend significant effort producing domestic propaganda and enforcing ideological orthodoxy, the distinction is blurry. For governments that obey laws prohibiting domestic propaganda and protecting freedom of expression, these activities may be very distinct realms of scholarship and activity.

[13] See Appendix E for sources and more detailed literature overview.

[14] See Appendix E for sources and more detailed literature overview.

process combines activity keywords with reaction language scanning, similar to some social media verification techniques.

Organization of This Report

Chapter Two describes the procedure we developed to extract chemical weapons incident information from social media using posts about the Syrian Civil War as our test case. Chapter Three describes implementation considerations for the procedure. Chapter Four concludes with a summary of findings and some brief remarks on how the work can support the National Defense Strategy (NDS) by enhancing situational awareness (part of the first line of effort in the NDS) and multiplying the capabilities of a social media analyst on this task by at least a factor of eight (part of third line of effort in the NDS).

In addition to the main body of this report, we have also included five appendixes in this publication, each of which provides more technical detail on various aspects of the approach.

CHAPTER TWO

Social Media Analysis of Chemical Weapons Incidents in Syria

In this chapter, we describe the procedure we developed to extract chemical weapons incident information from social media using posts about the Syrian Civil War as our test case. In general, the task is like finding a needle in a haystack, where the haystack is the voluminous quantity of social media content and the needle is content likely to be associated with a chemical incident. To find the needle, we need an approach for reducing the haystack to a hay*pile* small enough for a reasonably sized team of analysts to sort through in a reasonable amount of time. Then, we need an approach that the team can use to successfully distinguish needle from hay. The procedure draws on both computer and human strengths (see Figure 2.1).

This chapter provides basic instructions for each step of the procedure and demonstrates their application to chemical weapons use in Syria as a case study. This demonstration of the concept provides insight into the performance of the method for detecting chemical

Figure 2.1
Method Overview: Finding a Needle in a Haystack

Step 1	Step 2	Step 3	Step 4
Identify operationally relevant factors and examine known events to find incident indicators.	Develop a feed of social media data.	Conduct automated daily scans for elevated keyword use.	Analyze posts to verify detection and extract information.
Manually define features of the needle.	Computationally gather the haystack.	Computationally filter to smaller haypile.	Manually search for needle in the haypile.

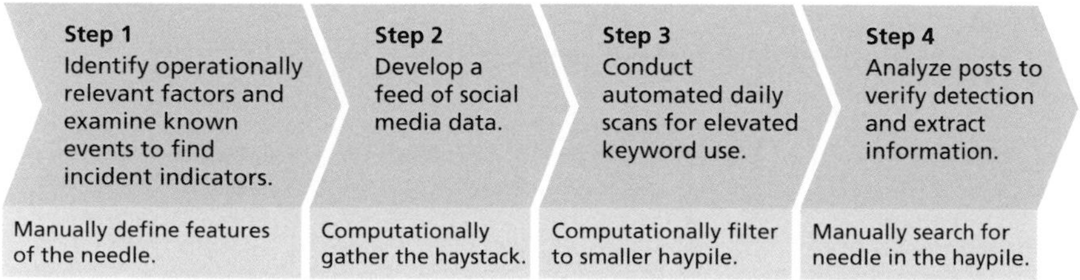

attacks and offers other potentially operationally relevant insights into their characteristics. For this demonstration, we rely on Twitter as our central data source for Steps 2 and 3. However, the methodology can be adapted to fit many social media platforms. Before discussing the four steps, we discuss precision and recall, which underlie this approach.

Precision and Recall: Understanding the Merits of Detection Methods

Throughout this chapter, we will refer to *precision* and *recall*, which are foundational measures of the quality of a detection procedure (a "detector"). A detector decides whether something is true or not true. The automated component of our procedure is a detector, which decides whether a chemical event has occurred (true) or has not (false). For a detector, there are four possible outcomes, because the detector can choose either true or false and because the correct answer can be either true or false. Table 2.1 shows each of those four outcomes and what they are typically called. For example, if the detector decides that the answer is false, and the right answer is actually true, that is called a *false negative* (FN).

Precision asks: "If the detector thinks that the answer is true, what are the chances that the correct answer is true?" Precision scores range from 0 to 100 percent, and equation 2.1 reports how they are calculated. Precision scores of at least 70 percent are generally considered good.

$$Precision = TP \; / \; (TP + FP) \qquad (2.1)$$

Recall asks: "If the correct answer is true, what are the changes that the detector thinks that the answer is true?" Recall scores range

Table 2.1
Four Possible Outcomes for a Detector Decision

	Correct Answer: TRUE	Correct Answer: FALSE
Detector decision: TRUE	True positive (TP)	False positive (FP)
Decider decision: FALSE	False negative (FN)	True negative (TN)

between 0 and 100 percent, and equation 2.2 reports how they are calculated. Recall scores of at least 70 percent are generally considered successful.

$$Recall = TP \;/\; (TP + FN) \qquad (2.2)$$

If our detector has high precision and low recall, it will not detect most chemical weapons events. However, it will be very credible when it does detect chemical weapons events. If our detector has low precision and high recall, it will detect most chemical weapons events. However, analysts will also receive many false alarms between each actual event.

It is difficult to make a method that has both high precision and high recall because there are trade-offs between the two: Many of the strategies that will increase performance on one measure will also decrease performance on the other.[1] Our strategy in developing this blended algorithmic-human procedure was to build an automated component with high recall that would filter the haystack down to a more manageable haypile and then trust our analysts to sort through that pile to find the true positives.

A consequence of this strategy is that there is also a trade-off between recall in the automated component and staff labor hours in the manual component. When we sacrifice precision to achieve better recall, we increase the staff labor costs necessary to sort through false positives to find true positives. However, the procedure will still lower labor costs significantly, relative to the cost of manually sorting through all the data without any automation.

[1] Cyril W. Cleverdon, "On the Inverse Relationship of Recall and Precision," *Journal of Documentation*, Vol. 28, No. 3, 1972; Michael Gordon and Manfred Kochen, "Recall-Precision Trade-Off: A Derivation," *Journal of the American Society for Information Science*, Vol. 40, No. 3, May 1989.

Step 1: Identify Operationally Relevant Factors and Examine Known Events to Determine Indicators of Incidents

The first step in our methodology requires us to *identify features of the needle*. *Features* are measurable aspects of social media posts that provide useful information. To identify features, we researched what information would be useful to decisionmakers and then determined how that information might appear in social media posts.

Identify Useful Information

We consulted with SMEs and reviewed documents from the U.S. Department of Defense, other U.S. government agencies, and public health sources to understand characteristics of a chemical incident that are useful for establishing battlefield situational awareness and elucidating threat networks. Table 2.2 lists the characteristics we identified as operationally relevant, which were determined after an extensive literature review and vetting with SMEs to confirm operational relevance.

Information about many but not all of these characteristics could plausibly be present in social media data. For example, it is important to determine both the geographical location of an incident and the manufacturing process used to produce chemical weapons. Manufacturers of such weapons, however, are unlikely to allow their work to be documented on social media, whereas it may be relatively straightforward to identify the location of an incident.

Characterize an Event as Reflected in Social Media

To understand how operationally relevant information might appear in social media, we reviewed 35 YouTube posts, 16 Twitter posts, and five Facebook posts that were associated with 19 previously documented events between 2012 and 2018. We also reviewed the comments and replies associated with each post—about 50 pieces of content per original post. Media and international observers, such as the Organization for the Prohibition of Chemical Weapons (OPCW),

Table 2.2
Operationally Relevant Characteristics of a Chemical Incident

Investigative Question	Characteristic
Who	• Attacker • Target • Other affected parties
When	• Time of day • Date
Where	• Geographic location • Meteorological conditions • Terrain • Proximity to critical infrastructure/hazard sites
What	• Signs that a violent incident took place • Chemical, biological, radiological, and nuclear (CBRN) versus kinetic
How	• Specialized equipment • Manufacturing

SOURCES: Donna Edwards, Paul Krauter, David Franco, and Mark Tucker, *Key Planning Factors: For Recovery from a Chemical Warfare Agent Incident, Summer 2019*, Washington, D.C.: U.S. Department of Homeland Security, Science and Technology, Summer 2012; Headquarters, Department of the Army, *Multi-Service Doctrine for Chemical, Biological, Radiological, and Nuclear Operations*, Washington, D.C., July 2011; Joint Publication 3-41, *Chemical, Biological, Radiological, and Nuclear Response*, Washington, D.C.: Joint Chiefs of Staff, September 9, 2016; U.S. Department of Defense, *The Militarily Critical Technologies List Part II: Weapons of Mass Destruction Technologies*, Washington, D.C.: Office of the Under Secretary of Defense for Acquisition and Technology, February 1998; Organization for the Prohibition of Chemical Weapons, "Syria and the OPCW," webpage, 2020; U.S. Department of Homeland Security, "Chemical Attack Fact Sheet: Warfare Agents, Industrial Chemicals, and Toxins," Washington, D.C.: National Academy of Sciences, 2004; World Health Organization, "Public Health Preparedness and Response," in *Public Health Response to Biological and Chemical Weapons: WHO Guidance*, 2nd ed., Geneva, 2004.

provided valuable insight about the evidentiary base for known events in Syria.[2]

---

[2] If no previous incident has occurred, analogous incidents might provide insight. For example, chemical weapons attacks may share similar features to chemical riot control measures or industrial accidents. For maximum utility, the search should be limited to incidents happening in similar cultural contexts, because language and social media behavioral norms are context-dependent.

In our analysis of known events, we considered two types of indicators of an event: (1) textual indicators and (2) imagery indicators. Although the textual indicators feed into the computational effort (Steps 1–3) that involve reducing the haystack to a haypile, the imagery indicators help the human analyst do his or her work to distinguish needle from hay (Step 4).

Identify Textual Indicators

In the known events we analyzed, we identified words and phrases (henceforth, *keywords*) included in tweets and text in associated retweets and comments that tended to appear in posts with relevant information, especially if the words were somewhat uncommon in a typical social media text. We identified two types of keywords: subject-matter words, such as chemical descriptors and symptoms, or reaction language, such as expressions of shock and grief.[3] Both sets of keywords are effective search terms, but they work for different reasons. To use an analogy, imagine trying to find an outdoor concert venue. Searching for subject-matter words is like listening for the sound of music to guide you there. Searching for reaction language is like listening for the roar of the crowd. Table 2.3 shows a selection of textual indicators we identified by analyzing known chemical events in Syria.

We identified a total of 41 subject-matter word candidates and 31 reaction-language candidates. Of them, nine subject-matter keywords and five reaction-language keywords demonstrated particularly strong potential—that is, they were neither too rare and technical for the average social media user (e.g., medical terminology) nor too common and generic to filter the haystack (e.g., common religious innovations). In our test data set of 56.6 million tweets (discussed in Step 2), 0.2 percent of tweets contained these subject-matter words, 0.6 percent contained reaction-language words, and 0.001 percent contained both. We next grouped these keywords into four keyword search groups: chemical names and symptoms, attack descriptors, angry expressions,

[3] Subject-matter language included names of chemical agents, symptoms of exposure, descriptors for a mass chemical incident (e.g., "toxic," "massacre"); reaction language included expressions of anger, reactions to tragic events, blame and/or accusatory language, and religious invocations.

Table 2.3
Selection of Textual Indicators Identified

Keyword Type	Keyword	Rough Translation	Significance
Subject matter	السارين	Sarin	Name of a chemical agent
	خروج الزبد من أفواه	Foaming at the mouth	Symptom of exposure
	مجزرة	Massacre	Descriptor of a mass attack
Reaction	الله ينتقم منك	May god take revenge on you	Expression of anger
	انا لله وانا اليه راجعون	We belong to god, and to god we shall return	Expression of mourning

and lamentations. The chemical names and symptoms and the attack descriptors groups consist of subject-matter keywords, and the angry expressions and the lamentations groups consist of reaction-language keywords. Figure 2.2 depicts the process of making keyword search groups.

In essence, this process replicates a form of rudimentary machine-learning modeling—selecting indicators, ruling out indicators that have low predictive power, and then compositing indicators to create stronger indicators. We offer no guidance on whether creating keyword groups manually or through machine learning is preferable. We use a manual process because we anticipate that—Syria not withstanding—chemical weapons attacks will generally be rare events. Human intelligence is better able to generalize from a small amount of data but doing so may introduce bias. Algorithms are better able to characterize statistical patterns, but they may have difficulty distinguishing event-specific keywords from general event type keywords in circumstances where few examples are available.

Identify Imagery Indicators

In addition to text, we also reviewed imagery and videos appearing in social media content associated with known chemical incidents in the test case. This review allowed us to understand how chemical incidents were reflected visually in social media and informed the development

Figure 2.2
Keyword Search Group Development Process

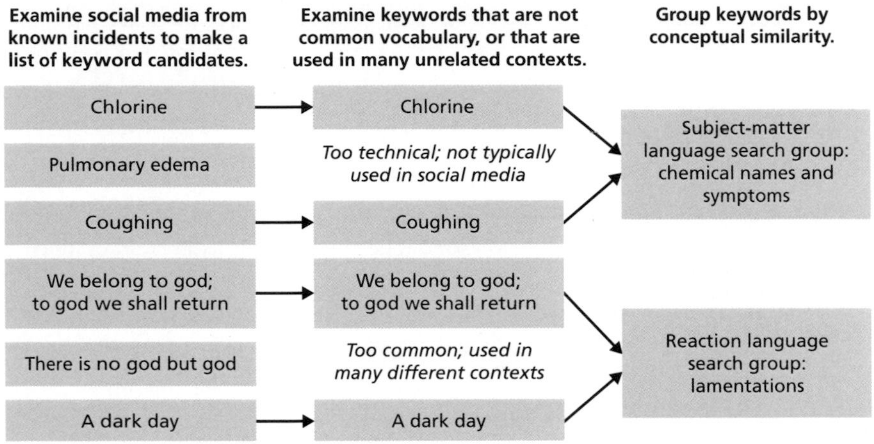

of a checklist to guide the analyst's review of social media content (Step 4).

Table 2.4 shows an excerpt from one of several checklists we developed—for different agents (chlorine, sulfur, mustard, and sarin), different points in time (scene of attack, hospital, and ex-post interviews), and different physical environments. Although some items were easily observed, others required more sophisticated inference. For example, identifying rust (images or video) or mentions of rust (text) at the attack site would be useful because chlorine—as opposed to sulfur, mustard, or sarin—will corrode metal.

We developed our checklists for this test case to help guide the analyst's structured review of social media content (Step 4), but more research is needed to refine the utility of the checklists. Table A.2 in Appendix A provides a notional example for how a more refined checklist might look.

Step 2: Develop a Feed of Social Media Data

In this step, we acquired a sample of the haystack large enough to contain needles. A key decision point in this step is which social

Table 2.4
Excerpt from Chemical Weapons Employment in Syria Checklists

Chlorine Exposure Indicator	What You Might See in Imager or Video
Distinctive odor	People mention the smell of chlorine; may describe it as smelling like "Klor" or "Flash," which are local brands for chlorine cleaning supplies
Decontamination procedures	Patients are hosed down prior to treatment; may appear damp but not from sweat
Symptoms of choking gas agents	Coughing, accessory muscle use in breathing, bloody sputum, wheezing/stridor, sneezing, pallor, blue tinged skin, gurgling noises, breathing sounds that suggest fluid in the lungs, productive cough
Checklist continues . . .	

media platform to use for automated data scanning. Social media characteristics in the region of interest, practical capabilities, and purpose should all inform this choice. Appendix B provides a detailed framework for assessing these considerations. As just noted, for this test case, we focused on Twitter.[4] Twitter's format is short, mobile-device oriented, and can require little bandwidth. This may facilitate posting from the scene of developing events, even in parts of the world where data infrastructure is limited. Twitter's permissive public access to data and text-centered medium make high-data volume and computerized analysis practical. In addition, its niche in the social media ecosystem tends to be oriented toward conversations about geopolitics and current events in a way that its larger competitors are not.[5] To develop our data set (the whole haystack), we built a list of active Twitter users, randomly sampled many users from it, and gathered every tweet those users had written after a given date. We discuss this process in the next section.

---

[4] For all parts of this process, it is important to note that Twitter has policies intended to restrict access to data and to discourage certain use cases. Thus, before using Twitter for monitoring, it is important to consult the terms of service to determine if a particular use case would be considered an acceptable use of the platform.

[5] The tagline on the sign-up page reads "See what's happening in the world right now," reflecting the company's current-events orientation.

Building the User List

The volume of social media posts generated worldwide is too large to be cost-effectively analyzed in real time. Sampling is a standard technique for remedying such situations. The first step to sampling is building a data set of users. The reason the sampling needs to be of users instead of posts is that a small minority of abnormally verbose users (often bots) generate the vast majority of tweets, and they are not representative of most users.[6] To build our list, we used the Twitter public stream over a period of months to accumulate a large list of active users who wrote tweets in Arabic.[7] Conducting simulation studies on the data gathered, we determined that it would likely require a minimum of 50 million tweets to build a representative list of users. We gathered significantly more than that for method-development purposes.

Randomly Sampling Users

Aiming to sample at least 200,000 users, we sampled slightly more than our goal and then removed from the sample any user who had made their tweets private or who had had their account suspended. After deletions, our sample consisted of 206,423 users. We provide no specific guidance on the optimal sample size because sample size is

[6] Bot and/or troll disinformation is a major concern in social media, and significant research effort is focused on it. See, for example, Matthew Benigni, Kenneth Joseph, and Kathleen M. Carley, "Bot-ivism: Assessing Information Manipulation in Social Media Using Network Analytics," in Nitin Agarwal, Nima Dokoohaki, and Serpil Tokdemir, eds., *Emerging Research Challenges and Opportunities in Social Network Analysis and Mining*, Cham, Switzerland: Springer, 2019; David Beskow and Kathleen M. Carley, "Bot Conversations Are Different: Leveraging Network Metrics for Bot Detection in Twitter," in *2018 IEEE/ACM International Conference on Advances in Social Networks Analysis and Mining (ASONAM) Proceedings*, Barcelona, Spain, August 28–31, 2018; Helmus et al., 2018.

[7] We do not recommend constraining this sample to the location of interest because geographic information on most social media platforms is unreliable and location constraints are not key to method performance. First, most users on Twitter do not provide accurate location information or geotag tweets. The remaining users tend to be unusual in ways that make them unrepresentative of the Twitter population at large. There are ways to infer geography for a larger segment of the population, but these violate the Twitter terms of service. Second, social media about chemical weapons incidents evokes strong reactions and gets widely shared as a result. Our method detects the wave of reaction sharing, thus alerting analysts to manually trace it back to its in-country source.

a trade-off between analyst staffing costs and recall—both of which increase as the sample size increases.

Gathering Data

We gathered every tweet that the sampled users wrote during the 2017–2018 period using Twitter's application programming interface (API) to test the procedure's effectiveness in sorting through the haystack on 730 separate days. We gathered a total of 803 million words from 56.6 million tweets. To use this method for continuous sampling, we recommend gathering the last one to two years of tweets to calculate the baseline frequency at which different keywords are used. Then, the API can be queried daily to collect each day's worth of tweets from sampled users to scan the data for signs of chemical weapons incidents.

Tamper Resistance

Organized, state-sponsored disinformation is a concern for any method that relies on social media data. The user-based sampling approach described here makes the method somewhat tamper-resistant, because disinformation agents would need to achieve a high degree of dominance over the information space to exert influence widespread enough to be seen in this kind of sample. We discuss the method's tamper-resilience features in Table A.4 in Appendix A.

Step 3: Conduct Automated Daily Scans for Elevated Keyword Use

In Step 1, we developed a list of keywords that users typically write when posting about incidents or reacting to those posts (see Table 2.3). We then combined those keywords into keyword search groups, where each group contains words that indicate the presence of social media posts that provide relevant information on chemical weapons incidents.

In Step 3, we counted how many users authored any tweets on each of the 730 days and then calculated the percentage of users who authored tweets using each of our four keyword groups on each day. We considered keyword group usage to be elevated if the percentile using any keyword in the group on that day was greater than the 98th percentile of keyword group usage across all 730 days. Mathematically, the 98th percentile of 730 is equivalent to treating the top 15 daily counts for each group as elevated and therefore noteworthy. Although we selected the 98th percentile as the noteworthiness threshold for our test run, we note that the optimal percentile will need to be calibrated to balance analyst staffing costs and recall. Both costs and recall increase as the percentile threshold decreases. It is also unknown whether a percentile-based threshold is superior to the scan-statistic approaches commonly used in epidemiological surveillance; both options merit consideration.

Performance of Computational Method at Detecting Known Events

To understand if elevated keyword use is an effective predictor, we used reporting from multiple news media, research, and nongovernmental organization sources to develop a composite list of 20 significant incidents in 2017 or 2018 that appeared in multiple sources (see Table A.1 in Appendix A).[8] We then attempted to predict on which days these incidents took place using our keyword groups. More formally, we treated all words in a keyword group as a single token, and if token usage was elevated on a day in our sample, we considered that keyword group to have predicted a chemical weapons incident on that day.

Table 2.5 reports the precision and recall scores these predictions achieved. We found that elevated usage of words in subject-matter keyword search groups made higher-precision, lower-recall predictions of chemical weapons incidents. That is to say, if subject-matter keyword usage was elevated on a given day, there was a 33 to 38 percent chance that one of our 20 documented incidents also occurred on that day (first two precision scores in Table 2.5). However, subject-matter keyword usage would only have brought 5 to 15 percent of the documented incidents to the attention of analysts (first two recall scores in Table 2.5).

---

[8] This list was made separately from the list used to identify keywords and indicators in Step 1.

Table 2.5
Performance of Keyword Search Groups as Predictors of Documented Chemical Weapons Incidents in Syria, 2017–2018

Group of Keywords	Precision	Recall
Subject-Matter Keyword Search Groups		
Names of agents or common symptoms	33%	5%
Descriptors of chemical weapons incidents	38%	15%
Names or symptoms OR descriptors	30%	15%
Reaction-Language Keyword Search Groups		
Angry reactions	6%	20%
Lamentations	6%	20%
Angry OR lamentations	7%	25%
All Keyword Search Groups		
Agent and symptoms OR descriptors OR angry expressions OR lamentations	10%	40%

In contrast, we found that elevated usage of words in reaction-language keyword search groups made lower-precision, higher-recall predictions of chemical weapons incidents. That is to say, if reaction language keyword usage was elevated on a particular day, there was only a 6 percent chance that one of our 20 documented incidents also occurred on that day (fourth and fifth precision scores in Table 2.5). However, reaction-language keyword usage would have brought 20 percent of the documented incidents to the attention of analysts (fourth and fifth recall scores).

For our procedure, recall is more important than precision because our human analysts can sift through false alarms (the result of low precision) but will have difficulty finding events that the algorithmic scan does not detect (the result of low recall). We blended subject-matter and reaction-language keywords together to boost recall—treating elevated keyword use for *any* of the four groups as evidence that an event has occurred. This produced our highest recall score—40 percent (bottom recall score in Table 2.5). In practical terms, this means that our detector would have been able to bring 40 percent of documented chemical weapons incidents to the attention of analysts (recall), but that analysts would experience about nine days of false alarms for each day when the procedure brought an actual chemical event to analysts' attention.

Typically, precision and recall scores of 70 percent or greater are considered signs that a detector is performing well. Despite this, the 40 percent scores in this proof of concept are promising for two reasons. First, achieving 40 percent recall with a small pilot like ours indicates that the methodology has potential and that performance could increase to higher levels with more development—more developed methodologies working on analogous problems have achieved well over 80 percent.[9] Second, our blended computational-human procedure provides a mechanism for coping with low precision scores: Analysts can sort through false alarms to find the true positives. A precision score of 10 percent means that analysts would have had to sort through social media on 80 days during the 2017–2018 period to find eight incidents. Compared with sorting through 730 days by hand, this is a significantly lighter workload.

Step 4: Analyze Posts to Verify Detection and Extract Information

Although the computational method (Steps 1–3) showed significant promise in terms of winnowing the haystack to a haypile, a human analyst is still needed to refine the detection of events by sorting needle from hay (Step 4). The output of the computational approach described in Step 3 would be an alert if total subject-matter or reaction-language keyword usage across all users was elevated on a given day, plus a list of

---

[9] Hyeok-Jun Choi and Cheong Hee Park, "Emerging Topic Detection in Twitter Stream Based on High Utility Pattern Mining," *Expert Systems with Applications*, Vol. 115, January 2019; Chao Fan and Ali Mostafavi, "A Graph-Based Method for Social Sensing of Infrastructure Disruptions in Disasters," *Computer-Aided Civil and Infrastructure Engineering*, Vol. 34, No. 12, May 2019.

users in the sample that used subject-matter keywords on the day for which the alert was issued.

In Step 4, the human analyst examined social media content in a structured manner. First, the analyst reviewed content from users who post using subject-matter keywords on days for which an alert was issued to determine if the alert was a false positive. Next, if the alert appeared to be a true positive, the analyst used a checklist to extract relevant information from posts that the user wrote that day. Finally, once all relevant information was extracted from the initial posts, the analyst vetted that information for accuracy and followed leads to extract additional information from more social media.

To understand whether this procedure was pointing analysts toward the right material, we conducted an additional test analysis of 49 posts that contained subject-matter and reaction-language keywords on days with or without elevated keyword use.[10] Table 2.6 reports our findings. The row indicates whether keyword usage was elevated for subject-matter keywords, reaction-language keywords, or neither on that day. The column indicates whether the post itself contained a subject-matter or reaction-language keyword. The percentages in each cell indicate the fraction of posts that contained useful information about a chemical weapons incident. We found that there was relevant information in 84 percent of posts that contained a subject-matter keyword and were written on days with elevated reaction-language and subject-matter keyword use (second column, bottom two rows of Table 2.6). In contrast, only 10 percent of the other posts in our test had relevant information. This difference is statistically significant at the 0.05 level and suggests that our procedure successfully guides analysts to needles in our haypile.

We used the checklist of indicators that we developed in Step 1 to extract additional information about the incident from posts. We found that more than half of posts with any relevant information

[10] We had originally planned for a larger sample size, but we became concerned about the effects that exposure to this content was having on our analyst and stopped the exercise at 49 posts. See Appendix D for guidance on protecting analysts from secondary posttraumatic stress.

Table 2.6
Analyst's Chances of Finding Chemical Attack Information, Given Elevated Keyword Use Across All Users and/or Keyword Use in Posts

Elevated Keyword Use That Day Among the Entire Sample of More than 200,000 Users	Keyword in Sampled User's Posts	
	Reaction Language (N = 23)	Subject-Matter Term (N = 26)
Neither (N = 17)	0%	29%
Reaction language (N = 13)	0%	83%
Subject-matter term (N = 19)	17%	85%
Elevated subject-matter or reaction-language keyword use that day AND subject-matter keyword in post (N = 19)	84%	
All others (N = 30)	10%	

contained information on the location of the event and the general kind of chemical agent involved. Some posts also contained information on the timing of the event and how the agent was deployed, but this type of information was much less common—present in roughly one in ten posts. Information on infrastructure, hazard sites, and other built structures was mostly absent. In reviewing each post manually, the most relevant information came from retweeted content that typically contained images, videos, and links. After analyzing linked content from just a handful of starting posts, we were generally able to gain a rough idea of the who, what, and where questions about the chemical weapons incident. Typical images or video focused on the victims, showing corpses and/or those receiving medical treatment. The condition of the victims provided evidence for determining if a chemical attack had occurred and with what agent. In terms of chemical versus kinetic attacks, chemical attack victims often did not have the same level of physical trauma as victims of kinetic attacks—fewer burns, less bleeding from obvious cuts, and the like. In terms of chemical agent, victims displayed distinct symptoms—for example, skin condition or color, respiratory secretions—that provided clues to the type of agent used. Although the checklists provided a structured analysis tool, they were developed for the narrow purposes of this proof of concept; future research would be required to refine, expand, and vet the checklists.

In addition to developing checklists to provide a structured approach for analysis of original content, we also compiled techniques for vetting social media and following leads outward from original content to gain additional information about user networks and information flows (Figure 2.3). We considered the post in the context of other user or online content and explored the original source for associated content, following links from the original post for additional clues. We also explored approaches to extract information from the imagery in the post itself by using distinguishing features of the terrain or physical environment to provide insight into the location. Table A.3 in Appendix A describes these concepts in more detail.

Conclusion on Performance of the Method on Test Case in Syria

In this chapter, we described the developed procedure and provided a demonstration of the concept using the Syrian Civil War. Although

Figure 2.3
Approaches to Vetting and Following Leads from Social Media Content

Check post information for consistency with other posts.

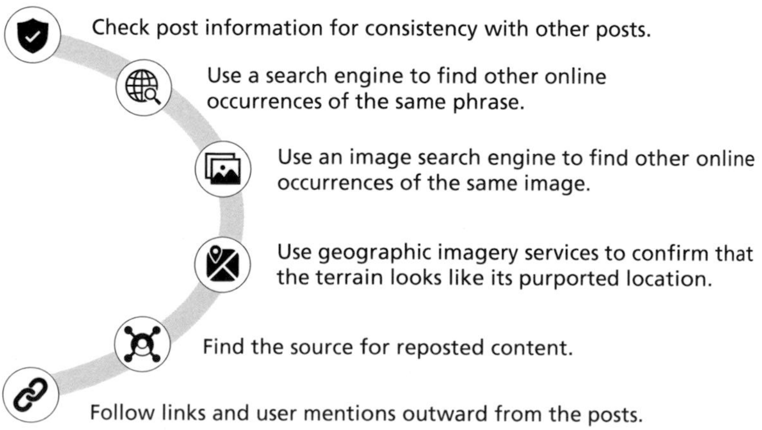

Use a search engine to find other online occurrences of the same phrase.

Use an image search engine to find other online occurrences of the same image.

Use geographic imagery services to confirm that the terrain looks like its purported location.

Find the source for reposted content.

Follow links and user mentions outward from the posts.

this small pilot study did not achieve operational levels of effectiveness, further refinement of the approach could yield the required performance.

In Step 1, we developed an indicator checklist based on recent chlorine and sarin use in Syria. This concept could be developed into a field manual for finding indicators for various chemical, biological, and radiological attacks in many cultural and geographic contexts.

In Steps 2 and 3, we developed an automated event-detection procedure centered on Twitter data that was able to alert analysts to 40 percent of documented events (i.e., recall). There are at least three opportunities to improve recall through further development. First, with more tweaks to the algorithm, recall could very likely be boosted to a much higher level. Second, the ground truth is uncertain in Syria, with different sources claiming that anywhere between 30 and 300 chemical weapons incidents took place. If the method was tested in a setting with clearer ground truth, recall could be more effectively evaluated. Third, if this procedure could be extended for use on imagery-centered platforms (e.g., YouTube) and on platforms that are key to specific geographic niches (e.g., Sina Weibo), it would increase the global reach of its capabilities.

In Step 4, we deployed a single analyst to the task of extracting relevant content from posts. An organized team of dedicated analysts would likely be able to achieve much higher levels of effectiveness and contribute to the development of a field manual for extracting and vetting social media information.

CHAPTER THREE

Implementation Considerations

To fully gauge this method's potential to become a tool of national security, we examined the practical and operational implications of our procedure. We determined that quality of the social media environment, staffing costs, and analyst secondary trauma risks are three factors that decisionmakers should consider if they decide to pursue this capability. These considerations are discussed briefly here, with additional material provided in Appendixes B–D.

Evaluate the Quality of the Social Media Environment

Information gathering from social media is only an effective option in regions of the world with a social media environment that is conducive to analysis. For example, the population of potential social media users must have reliable access to electricity, the internet, and devices, and these services need to be available at price points low enough that non-elite members of society can afford them. The population also needs to be literate in one of the languages commonly supported on the internet and tech-savvy enough to post and interact with social media content.[1]

[1] Katharine Schwab, "The Internet Isn't Available in Most Languages," *The Atlantic*, November 2015. For example, Facebook supports roughly 110 languages, YouTube supports about 80, and Wikipedia has at least 1,000 articles to offer for approximately 250 languages. For reference, Ethnologue estimates that 88 percent of the world's population speak one of the most common 200 languages, and 64 percent speak one of the top 20 languages (David M. Eberhard, Gary F. Simons, and Charles D. Fennig, eds., "Ethnologue: Languages of the World," 23rd ed., website, Dallas, Tex.: SIL International, undated).

In addition, a nontrivial percentage of the population needs to regularly use that area's most common social media platforms. Governance factors are also significant, because state-sponsored censorship and/or disinformation can potentially degrade the information value of social media. Platform policies are also critical to consider, because social media platforms vary in permissiveness toward bulk data collection and specific uses of those data.

Besides these contextual features of the information environment, the utility of social media analysis is also a function of analysts' capabilities. They must understand regional languages and be culturally fluent enough to understand the idioms, abbreviations, and references in the internet dialect (chatspeak) version of them. Analysts also must have the capability to collect data from the unique platforms popular in a particular region, which may differ from what is commonly available in the West. In addition, fact-checking locations requires geographic databases with a rich store of imagery for the purported location.

Syria includes some factors associated with a positive social media environment and some associated with a more challenging environment. In 2019, 33 percent of the population used the internet, and 31 percent used a mobile device to access it. Reported use of social media was actually higher than internet usage, with an estimated 37 percent describing themselves as active. Platform-specific social media usage data are difficult to obtain, but Twitter.com was the 15th most-visited website in Syria in 2019, suggesting that its use is relatively widespread.[2]

However, Syria routinely ranks among the worst countries in the world for suppression of information and other human rights. The Syrian government engages in extensive surveillance of online content, and it persecutes activists, censors content, and cuts off internet access in response to unfavorable activity.[3] In this degraded environment,

---

[2] DataReportal, "Digital 2019: Syria," DataReportal webpage, January 31, 2019.

[3] State-sponsored pro-regime hacking groups, such as the Syrian Electronic Army, have launched cyberattacks on activists, news media outlets, and messaging apps. See Abdelberi Chaabane, Terence Chen, Mathieu Cunche, Emiliano De Cristofaro, Arik Friedman, and Mohamed Ali Kaafar, "Censorship in the Wild: Analyzing Internet Filtering in Syria," *IMC '14: Proceedings of the 2014 Conference on Internet Measurement Conference*, New

social media use is both more challenging and more important for establishing battlefield situational awareness. Appendix B provides a 12-point framework for assessing the local social media environment.

Estimating Analyst Level of Effort Requirements for This Analysis

Another critical implementation consideration is the level of effort required for analysts to review the haystack of social media content; thus, we have estimated the potential manpower requirements for this method. To develop this estimate, we timed analysts in reviewing each unit of social media content and made assumptions as necessary to extrapolate to a number of units reviewed in a typical year.[4] We found that analysts needed three to eight minutes to assess one unit of content that contained no relevant information and five to ten minutes to assess one unit of content that did contain relevant information.[5] Posts with relevant content took longer because the analyst playbook calls for cross-checking post information against other sources and, for posts with links, examining the media at the other end of those links.

Extrapolating from these ranges per unit of content, we estimate that it would take the staff time equivalent (STE) of 70 to 230 full-time analysts to manually analyze social media from all sampled users without using our computational procedure to narrow down the haystack. The 70 STE estimate assumes that, over the long term, analysts will need an average of three minutes to analyze a unit of content without relevant information and five minutes to analyze a unit of content with relevant information. The 230 STE estimate assumes

---

York: Association for Computing Machinery, 2014; Zack Whittaker, "Surveillance and Censorship: Inside Syria's Internet," *CBS News*, December 12, 2013.

[4] The unit of analysis was a *user-day*: all tweets on a given user's account on a given day. We assumed seven hours a day of work on this analysis and 230 workdays a year.

[5] Ranges represent the 10th and 90th percentiles of time expended. STE estimates cover one language only.

eight minutes for a unit of content without relevant information and ten minutes for a unit with relevant information.

As shown in Table 3.1, using the computational procedure reduces STE requirements by 95 percent, resulting in an estimated workload of 1–29 STE. Most of this workload would consist of analysts waiting on standby because no work would be needed on days with no keyword surge.

However, the staffing requirement can be reduced if analysts' costs are only incurred on-demand, when surges in keyword use are detected. Doing this would allow staff to be allocated to social media analysis on an as-needed basis. Perhaps these staff conducted a different kind of analysis that requires similar skills and can be rapidly re-tasked if the algorithm detects elevated keyword use. Although peak STE usage would still require 1–29 STE, total annual STE usage would be much lower. Assuming this kind of rapid re-tasking, a sample that is similar to our test case data, and a 230-day work year, our social media protocol could only require the cost equivalent of 3–5 STE. Appendix C provides a more detailed discussion of the workforce estimates.

Protecting Analysts from Secondary Trauma

If an organization decides to implement a social media analysis capability for counter–weapons of mass destruction or other missions, it is also important to mitigate risks to analysts' mental health and readiness. Social media content can include disturbing text, images,

Table 3.1
Time to Complete Manual Analysis of Posts and Linked Media

Requirement	Analyst workload required to review haystack every day (without computational help)	Analyst workload required to be on standby to review haypile whenever a keyword surge is detected (with computational help)	Analyst workload required to review haypile only on keyword surge days in a given year
STE	70–30	1–29	3–5

and videos depicting violence, suffering, and physical destruction; this was certainly the case for our analysis of chemical weapons use. Given this, a decisionmaker needs to be aware of the risk of vicarious traumatization or secondary traumatic stress to the analysts tasked with conducting this work. These two concepts refer to the impact of indirect trauma exposure on professionals, such as researchers and analysts, mental health practitioners, educators, journalists, and human rights activists. Thus, implementing a social media–analysis capability should take into account both the potential deleterious health and safety effects on analysts and the secondary trauma, which can limit the long-term readiness, including the efficiency or effectiveness, of analysts. To support readiness objectives, Appendix C provides an additional discussion of secondary trauma drawn from past RAND Corporation work, including risk factors, symptoms, self-care best practices, and references.

Conclusion About Implementation Considerations

Although social media analysis appears promising as a way to provide battlefield situational awareness of the use of chemical weapons, decisionmakers considering implementation should proceed carefully. The suitability of such analytic techniques is a function of regionally specific features and available resources. Furthermore, analysis of potentially disturbing content can take an emotional toll on personnel, so steps should be taken to mitigate risks.

CHAPTER FOUR

Conclusions and Policy Implications

This research project developed an approach for analyzing social media data to derive insights about chemical weapons incidents and applied the approach to Syria as a proof of concept. We found that social media can provide low-latency, warfighter-relevant insights. The approach developed, which blends automated and human analytics, is more effective than either on its own. Our findings are summarized below as answers to the four research questions posed in Chapter One:

- **When a chemical event happens, what information about it tends to appear in social media, and what forms does that information typically take?** We find that Twitter, Facebook, and YouTube post text may include chemical attack descriptor keywords, place names, and broad characterizations of the agent used. It may also include reactive language, such as expressions of anger, lamentation, or religious invocation. Images, videos, and links may provide more incident detail, as well as ways to verify authenticity.
- **How can a computational approach enable us to rapidly detect chemical weapons incidents buried among millions of social media posts?** Scanning a Twitter sample for chemical incident–descriptor keyword and reactive-language surges can provide analysts with a low-latency alert that an event may have happened. This can direct them toward posts they can use to adjudicate whether it has happened. We scanned Twitter during our research for this report, but choosing which platform to scan is a complex decision with many considerations.

- **How can a blended computer-human approach improve on a fully automated approach?** Human intelligence may be better able to cope with the irregularity of social media data and better poised to use sophisticated inference and verification techniques to investigate incidents. If computation can filter data volume down to something manageable, the human intelligence component can generate deeper insights from it.
- **What do leaders need to know to implement a social media–analysis capability?** The information value of social media methods varies from place to place, depending on characteristics of the population, relevant state actors, social media platforms, and analyst capabilities. Implementing this method also requires supporting proportional staffing levels and taking steps to protect staff from posttraumatic stress.

DTRA commissioned this work as part of its ongoing mission to deter and counter weapons of mass destruction and support the NDS.[1] Our findings contribute to that mission in three ways. First, NDS line of effort one strives to build a more lethal force by supporting situational awareness. This method exploits a low-latency data source to provide rapid visibility on chemical weapons incidents and may be effective in denied environments in which alternative sources have been rendered ineffective. Second, NDS line of effort three strives to reform the department for greater performance and affordability. This method uses a blended computer-human approach that enables social media analysts to derive timely insight from the massive data streams of social media and empowers one analyst to do the work of eight. Third, the NDS identifies big-data analytics as a rapidly evolving technology that can "ensure we will be able to fight and win . . . wars" and asserts that "the Joint Force must gain and maintain information superiority."[2]

---

[1] House Armed Services Committee, *Statement of Vayl Oxford*, testimony before the Subcommittee on Intelligence and Emerging Threats and Capabilities, Washington, D.C., February 11, 2020.

[2] Jim Mattis, *Summary of the 2018 National Defense Strategy of the United States of America*, Washington, D.C.: U.S. Department of Defense, 2018, pp. 3, 6.

This method offers a tool for quickly generating information from large, unstructured data. Although this demonstration of concept study has focused on chemical weapons incidents, the method has many potential applications that would support the NDS.

Recommendations for Moving Forward: Research and Development Phase II

This demonstration of concept suggests that a blended computer-human procedure can sift through unstructured social media data to detect emerging chemical weapons events and extract relevant information. However, this demonstration only indicates that the approach is possible, without adjudicating about whether it is practical. Thus, a second phase of research and development is necessary to lay the ground for this program.

We recommend that DTRA initiates three activities to improve cost-effectiveness, improve event-detection effectiveness, and build practical experience. These activities would have two goals. First, the activities would thoroughly evaluate whether the method is effective enough to be stood up as a new capability, culminating in a recommendation to proceed or not. Second, the activities would build and mobilize the resources needed to stand up a program that implements this capability. This may enable the new program, if green-lit, to become operational quickly and smoothly. The subsequent paragraphs describe each activity. Table 4.1 summarizes the recommended activities, which are discussed in more detail in the next sections.

Recommendation #1: Develop Resource-Pooling Agreements with Other Federal Agencies

As with any new capability, cost-effectiveness is an important consideration. Our experiences suggest that the two most significant costs may be maintaining culturally competent staff to institute the manual analysis component and maintaining a data-science team to institute the computational component. We anticipate culturally

Table 4.1
Recommended Phase II Research and Development Activities

Recommended Activity	Evaluation Questions	Resource for Program Stand-Up
#1: Consider developing resource-pooling agreements with other federal agencies.	• Can pooling agreements lower the cost of instituting this capability? • Can pooling agreements ensure an adequate supply of cultural expertise?	• Ability to surge cultural specialists across multiple federal agencies when a relevant emerging event is detected • Cost-sharing agreement among multiple federal agencies to support the information technology or data science team implementing the method's computational part
#2: Consider building an extensive chemical weapons keyword list and rigorously test it against best available intelligence.	• How effective can a chemical weapons incident detection scan become?	• Extensive, validated keyword list for chemical weapons • Stronger evidence of effectiveness for the chemical weapons test case
#3: Consider conducting an exercise with federal computers and staff, scanning for easily verified types of events.	• What implementation challenges would this capability face and are there effective solutions? • How effective might a multiagency capability be across a range of verifiable event types? • How likely is it that a pilot program would be cost-effective?	• Staff experienced in conducting all phases of the process • Standard operating procedure and training documents • Evidence of effectiveness across multiple test cases

competent staff to be the larger cost challenge for two reasons. First, cultural competency is a highly specialized skill. Depending on mission requirements, the program might need dozens, if not hundreds, of different cultural specialties to cover all cultural groups of interest. Second, the method requires an analyst surge on a small number of days, rather than requiring a consistent amount of analyst labor each day. We anticipate the information technology and/or data science team being a cost challenge as well, because there is strong market demand for the required technical skills.

For both cases, resource pooling may be a viable cost-control approach. Chemical weapons incidents are only one potential application of this method. Multiple federal agencies could benefit from applying this method to meet their needs.[3] A joint program could use a common set of resources to meet these needs, thus defraying costs. In terms of cultural competency, many federal agencies conduct open-source analysis and thus require culturally fluent analysts. A pooling agreement would allow these analysts to be temporarily re-tasked to sort through social media on days when the computerized scan detects a surge in relevant keywords. In terms of information technology and/or data science, only one team is needed to institute the computational component of this method across cultures and use cases. Multiple federal agencies could split the costs of this team.

Recommendation #2: Build an Extensive Chemical Weapons Keyword List and Rigorously Test It Against Best Available Intelligence

The demonstration of concept discussed in this report scanned for potential events using an algorithmic procedure to find indicative keyword surges. Despite being a rudimentary exploratory effort, the procedure was able to find 40 percent of events during the test period. However, larger efforts have achieved higher detection rates on rated detection problems. These efforts suggest that there are two ways the procedure can be developed further, thus potentially improving the detection rate into ranges suitable for operational use. First, the keyword list could be expanded significantly, especially in terms of better capturing the nuances of reaction language in cultural content. A larger, more nuanced list would likely be more effective at detecting surges. Second, the list of test events could be based on the best available intelligence, instead of open-source reporting. Available open-source reporting was vague and sometimes offered conflicting information.

---

[3] More specifically, the U.S. Department of Defense, intelligence community, and the U.S. Department of Homeland Security.

A list with reliable dates and casualty figures would make detection calibration more effective and accuracy testing more precise. With these two improvements, a second round of testing and calibration may achieve much higher levels of detection effectiveness.

Recommendation #3: Conduct an Exercise with Federal Computers and Staff, Scanning for Easily Verified Types of Events

Before this method can be implemented as a new capability, it must be translated into procedural documents, and program staff trainers need to master the method. In addition, the multiagency funding case for this capability would be stronger if it was tested against multiple types of events under multiple conditions, especially since the cost of testing for multiple event types in the same region is not much higher than the cost of testing for a single event type.[4] These activities bridge the gap between a promising basic research finding and a shovel-ready pilot program.

Conduct an exercise to build that bridge. In this exercise, federal staff and computers would be used to detect and analyze easily verified events, such as recent natural disasters. Researchers would design the exercise, improve procedural documentation, and conduct after-action analysis. The resulting data would provide credible estimates of whether a pilot program would be effective, cost-effective, and feasible.

Combined, the recommended activities provide all the information necessary for a decisionmaker to determine if a pilot program is warranted, including effectiveness, cost-effectiveness, and feasibility. In terms of cost-effectiveness, recommendation #3 would enable researchers to estimate operational costs without having to incur the full burden of those costs in the process. Recommendation #1 would significantly improve the efficiency with which staff resources are used and would diffuse costs. In terms of effectiveness, recommendation #2 would

---

[4] Most components of the methodology are the same regardless of use case. Each new event type would require a new subject-matter keyword list and analyst checklist. However, the event-detection computer program, analyst playbook, reaction-language keyword list, and requisite cultural expertise are the same for any region of interest. This makes the cost of testing against multiple use cases not much higher than the cost of testing against a single use case in that region.

improve and supply evidence of effectiveness for the computational scan component. Recommendation #3 would demonstrate method effectiveness across a wide array of test cases. In terms of feasibility, recommendations #2 and #3 generate all the procedural documents and resources needed to launch a pilot program, as well as anticipate any challenges that a pilot might face. They also supply credible evidence that the pilot program will succeed. Recommendation #1 ensures the scarcest resource—cultural expertise.

APPENDIX A

Additional Technical Tables

This appendix contains some additional technical tables that are noted in the main text.

Documented Chemical Weapons Incidents in Syria, 2017–2018

Table A.1 lists 20 chemical weapons incidents that we considered ground truth in evaluating the performance of our computational method. Reports vary widely, with some sources listing only a dozen incidents and others reporting hundreds. In addition, incidents varied in casualty count and public visibility, both of which affect how easily our procedure can detect them. To make our list in Table A.1, we compiled reporting of multiple sources and then cross-checked for incidents that appeared in many of them.

Notional Checklist

Table A.2 depicts a notional example of how a more developed analyst checklist might look. The checklist is organized around the five classic investigative questions (first column), breaking down each into categories of information (Feature) that are operationally relevant for our purposes. The next column (Observable Indicators) lists examples of specific observations that could be a sign of one of the features. The final column (Analyst's Observation) provides space for the analyst to write down relevant observations.

Table A.1
Documented Events

Date	Location	Chemical
2017 Jan 07	Bassimeh	Chlorine
2017 Jan 29	Sultan Al-Marj	Chlorine
2017 Mar 23	Latamenah	Sarin
2017 Mar 24	Latamenah	Chlorine
2017 Mar 28	Qaboun, Damascus	Chlorine
2017 Mar 29	Latamenah	Mixed reporting
2017 Apr 03	Khan Shaykhun	Mixed reporting
2017 Apr 06	Qaboun, Damascus	Chlorine
2017 Jun 30	Ein Tarma	Chlorine
2017 Jul 01	Zamalka, Damascus	Chlorine
2017 Jul 05	Jobar, Damascus	Chlorine
2018 Jan 12	Douma	Chlorine
2018 Jan 21	Douma	Chlorine
2018 Jan 31	Douma	Chlorine
2018 Feb 04	Idlib	Chlorine
2018 Feb 15	Afrin	unknown
2018 Feb 25	East Ghouta	Chlorine
2018 Mar 06	East Ghouta	Chlorine
2018 Apr 06	Douma	Mixed reporting
2018 Nov 24	Aleppo	Chlorine

SOURCES: ABC News, "Russia and Syria Blame Rebels for Suspected Poison Gas Attack in Aleppo," November 25, 2018; Sarah Almukhtar, "Most Chemical Attacks in Syria Get Little Attention. Here Are 34 Confirmed Cases," *New York Times*, April 13, 2018; Associated Press, "Timeline of Chemical Weapons Attacks in Syria," April 10, 2018; Anne Barnard and Hwaida Saad, "Despite U.N. Cease-Fire, Syrian Forces Begin New Attacks in Rebel-Held Enclave," *New York Times*, February 25, 2018; BBC News, "Syria War: 'Chlorine Attack' on Rebel-Held Idlib Town," webpage, February 5, 2018; Mike Corder, "Watchdog: Chlorine Used in Syrian Town of Saraqeb," Associated Press, May 16, 2018; Eliot Higgins, Images from the January 22nd 2018 Chlorine Attack in Douma, Damascus," *Bellingcat* blog, February 5, 2018; Ben Hubbard,

Table A.1—Continued

"Syria Urges U.N. to Condemn Rebels After Apparent Chemical Attack," *New York Times*, November 25, 2018; Ajay Nair, "Chemical Attacks in Syria: A Deadly History," *Sky News*, April 14, 2018; Reuters, "Rescuers in Rebel-Held Syrian Area Accuse Government of Gas Attack," January 22, 2018; Reuters, "Turkish Army Hit Village in Syria's Afrin with Suspected Gas: Kurdish YPG, Observatory," February 16, 2018; Alicia Sanders-Zakre, "Timeline of Syrian Chemical Weapons Activity, 2012–2019," Arms Control Association website, March 2019; Sarah Sirgany and Eyad Kourdi, "Dozens Injured in Toxic Gas Attack on Aleppo, Syria, Reports Say," CNN, November 26, 2018; Schneider and Lütkefend, 2019; Umut Uras, "Turkey Official Denies Use of Chemical Weapons in Afrin," *Al Jazeera*, February 18, 2018; United Nations Office of the High Commissioner on Human Rights, "Chemical Weapons Attacks Documented by the Independent International Commission of Inquiry on the Syrian Arabic Republic," January 15, 2018.

Analyst Tactics for Following Leads and Vetting Information

Table A.3 describes tactics that analysts can use to follow leads outward from keyword-bearing posts and vet the information found.

Tamper Resistance

The term *disinformation* is defined as organized campaigns convincing a population that something is not true. Disinformation campaigns may be state-sponsored. *Misinformation* is the term for the spread of incorrect information, often as a result of confirmation bias and social homophily. Social media is vulnerable to disinformation and misinformation. In countering these risks, our procedure incorporates four tamper-resistant features, described in Table A.4.

Table A.2
Notional Checklist

Investigative Question	Feature	Observable Indicators	Analyst's Observation
Who	• Attacker • Target • Other affect parties	• Uniforms • Language • Distinctive equipment or brands • Nearby populations	*Carrying a Russian brand of cigarettes. Office building nearby.*
When	• Time of day • Date	• Darkness; position sun or moon • Seasonal indicators • Date stamps and/or metadata • Date-apparent advertisements	*Snow on the ground. Sun is close to horizon–sunrise or sunset. Billboard references upcoming holiday.*
Where	• Geographic location • Meteorological conditions • Terrain • Proximity to critical infrastructure or hazard sites	• Recognizable landmarks • Signage • Visible weather (e.g., wind, rain) • Clothing (implies weather) • Visible terrain	*People are wearing heavy coats and the wind is moving their hair. Store sign in background (has six locations in the area).*
What	• Signs that a violent incident took place • Chemical, biological, radiological, and nuclear versus kinetic	• Injuries consistent with projectiles • Explosions • Symptoms consistent with specific agent	*Bodies on the ground with shrapnel wounds.*
How	• Specialized equipment • Manufacturing	• Vehicles • Weapons • Creation of improvised items	*Sound of helicopter in background. Parts for making IED [improvised explosive device] on table.*

Table A.3
Analyst Tactics for Following Leads and Vetting Information

Tactic	Description
Compare posts with each other.	Check post information for consistency with other posts. If posts from many users agree on certain information, that information is more likely to be accurate. When checking for consistency, make sure that the information is not consistent as the result of all users getting the information from a single common source.
Search for identical phrasing.	Use a search engine (e.g., Google.com, DuckDuckGo.com, Bing.com) to find other places on the internet where nearly the exact same phrasing is used and evaluate how that affects the credibility of the information. If a group of users all post near-identical phrasing, representing it as their own, the information is likely part of an influence campaign—not credible.
Search for similar ideas and evaluate website credibility.	Use a search engine (e.g., Google.com, DuckDuckGo.com, Bing.com) to find other places on the internet that reference the same ideas and make note of the credibility of those websites. If the top search results are conspiracy websites or known propaganda outlets, the post may not be credible.
Search for image sources.	Use an image search engine (e.g., Tineye.com, google.com/imghp) to find other places on the internet where the same images may be posted. A common disinformation technique involves misrepresenting old images as footage of current events.
Compare geography with geographic image databases.	Use geographic imagery services (e.g., google.com/maps Street View, openstreetcam.org, bing.com/maps StreetSide) to confirm that the terrain in images or video recordings looks like its purported location. In parts of the world where street imagery is unavailable, photo-sharing sites (e.g., flickr.com, the Instagram app) may be an alternative. When trying to identify locations, visible signage is often the quickest way to narrow down where to look: If the business's name is known, its location can be found (via google.com/maps, yelp.com, the website for the business).
Trace reposted content back to source.	Find the source for reposted content. Many posts repeat content from other users. This is called "reposting" or "retweeting." For these posts, find the original post and evaluate its credibility. If the source account has been suspended or does not read like a real person manages the account, the information may not be credible. Signs that an account is fake include most posts being incoherent, make no reference to typical life activities or events, consist of rehashed content (e.g., song lyrics), or are grammatically inconsistent with the purported native language of the speaker.

Table A.3—Continued

Tactic	Description
Trace a user's influences.	Follow links and user mentions outward from the posts. This can reveal more social media that may be relevant, especially social media written on other platforms. In addition, this may shed light on what or who influences the user. When following links and mentions, strive to only visit established social media sites—less reputable websites may infect a computer with viruses. Subject all posts found through this process to the same vetting tactics described above.

Table A.4
Tamper-Resistant Features of Our Procedure

Feature	Tamper-Resistant Value
User-based keyword elevation monitoring	We sample users and monitor for elevated keyword use across those users. This is more tamper-resistant than monitoring directly for trending keywords because concerted bot or troll activity can trick trending algorithms more easily than they can influence millions of individual users to tweet in a specific way on a specific date. Because the trolls or bots do not know which users we are monitoring, they would need to mount a truly pervasive campaign to influence the sample.
Verbosity-adjusted user sampling	Most tweets come from a minority of highly verbose accounts, which are qualitatively different than the typical user. Such accounts may be team-managed and/or use bot automation in an attempt to dominate a social media space through volume. In sampling users, we calculate each user's selection chance as the inverse of their lifetime daily rate of tweet production. This makes the resulting user sample representative of the user population and limits the chances that high-volume accounts become part of the sample.
Mentioned-adjusted keyword elevation counting	We scan each user's past tweets for mentions and consider users to be linked if they mention any of the same users. For the purpose of determining if keyword usage is elevated, we discount keyword use from linked users because we expect their keyword use to not be statistically independent of each other. This prevents our keyword-elevation detection procedure from being unduly influenced by rumors spreading between users in our sample.
Source triangulation	Our analyst playbook includes a variety of tactics for confirming information by cross-checking it against other sources, including geographic databases. This screens out many of the most common forms of fabricated information.

APPENDIX B

Quality of the Social Media Environment

In Chapter Three, we discussed various considerations to take into account if policymakers are thinking about implementing the social media–based methods discussed in the report. In this appendix, we expand on the discussion in Chapter Three in terms of the quality of the social media environment in a specific geographical area.

Four Factors That Need to Be Considered

When determining whether social media–based methods will be effective in a specific geographic area, at least four major factors should be considered: population, governance, platform policy, and analysis capability.

- **Population:** A sizable fraction of the population uses broadcast-oriented social media regularly, and there are no technological or socioeconomic barriers to effective use of it.
- **Governance:** State-sanctioned actors are not able to deter posting and post-sharing of relevant material or obscure it through disinformation.
- **Platform policy:** Usage policy for suitable social media platforms does not prohibit an organization from employing this methodology, nor does it limit data transfer in ways that render it ineffective.

- **Analysis capability:** Analysts are culturally fluent, technologically supported, and have access to sufficient regional database material to facilitate cross-checking.

Tables B.1–B.4 in this appendix describe the population, governance, platform policy, and analysis capability considerations in detail.

Detailed Breakdown on the Four Factors

Population Factors

Social media–based methods are most likely to be effective when a sizable fraction of the population uses broadcast-oriented social media

Table B.1
Population Factors

Population Factors	Notes
A sizable fraction of the local population has some access to electricity. Find data here: World Bank, "Access to Electricity (% of Population)," webpage, undated, https://data.worldbank.org/indicator/EG.ELC.ACCS.ZS	Periodic access to electricity is crucial to powering devices for accessing social media platforms and capturing images and video. Widespread access to fuel and portable generators may fill this requirement in some locations. If access is limited to just the wealthy, proceed with caution—elite perspectives and narratives may differ markedly from the experiences of the population at large.
A sizable fraction of the local population uses the internet regularly. Find data here: World Bank, "Individuals Using the Internet (% of Population)," webpage, undated, https://data.worldbank.org/indicator/IT.NET.USER.ZS	Periodic access to the internet is crucial for accessing social media platforms. Internet access takes different forms in different countries—landline, mobile device access, and/or internet cafes; thus, it is better to measure the portion that regularly use the internet and not the method of access. Regular use is also important because it takes practice to learn how to do sophisticated things, such as disseminate eyewitness footage through social media platforms. If access is limited to just the wealthy, proceed with caution—elite perspectives and narratives may differ markedly from the experiences of the population at large.

Table B.1—Continued

Population Factors	Notes
A sizable fraction of the local population has technical devices suitable for social media access. Find data here: World Bank, "Mobile Cellular Subscriptions (per 100 People)," webpage, undated, https://data.worldbank.org/indicator/IT.CEL.SETS.P2 We Are Social, "Digital in 2020," webpage, undated, https://wearesocial.com/digital-2020	Access to technology is crucial for accessing social media platforms and capturing images and video. Widespread access to affordable smart phones is a particularly good sign for this method, because such devices provide both platform access and images and video capture. If access is limited to just the wealthy, proceed with caution—elite perspectives and narratives may differ markedly from the experiences of the population at large.
A sizable fraction of the local population uses broadcast-oriented social media. Find data here: We Are Social, "Digital in 2020," webpage, undated, https://wearesocial.com/digital-2020. These presentations are updated annually, so next year's reports will be called "Digital in 2021," and so forth.	There are at least two major types of social media—personal communication–oriented and broadcast-oriented. Personal communication–oriented platforms, such as WhatsApp and Telegram, provide a method of intentional communication between acquainted persons. Broadcast platforms, such as Twitter and TikTok, provide a method to disseminate user-generated content and typically encourage dissemination beyond a user's acquaintances. Most broadcast platforms also support personal communication through direct messaging features. For example, the Explore and Hashtag features of Instagram are broadcast-oriented, but the platform also supports direct messaging.

regularly, and there are no technological or socioeconomic barriers to effective use of it. Table B.1 discusses these factors in detail.

Governance Factors

Social media–based methods are most likely to be effective when state-sanctioned actors are not able to deter posting and post-sharing of relevant material nor obscure it through disinformation. Table B.2 discusses these factors in detail.

Table B.2
Governance Factors

Governance Factors	Notes
State-sponsored censorship of social media in this locale is minimal or ineffective. Find data here: Freedom House, "Internet Freedom Status," webpage, undated, https://freedomhouse.org/explore-the-map?type=fotn&year=2019 Reporters Without Borders, "World Press Freedom Index: Ranking 2020," webpage, 2020, https://rsf.org/en/ranking	State-sponsored censorship of social media occurs in many countries. However, censorship efforts vary in extent and sophistication. Basic approaches include shutting down the internet during times of unrest, regulating access to the internet in ways that diminish anonymity, and persecuting high-profile social media activists that come to the attention of authorities or state-sanctioned groups. More sophisticated approaches might include real-time content filtering, censor review of social media as it is being posted, and extensive domestic influence campaigns. In addition, strong state pressure can lead to self-censorship that extends beyond direct state intervention. If social media within a state is subject to extensive, well-resourced censorship, consider conducting testing or retrospective analysis to determine if informative social media is likely to be or remain published. When conducting such research, consider these research questions: Are activists typically successful in publishing exposé or current event social media content? Are common users typically successful in sharing expose or current event social media content?
No external state has a strong interest in suppressing or distorting this kind of information in this locale. Find data here: Various analyses of information on MediaWell webpage, https://mediawell.ssrc.org Bradshaw and Howard, 2019, https://comprop.oii.ox.ac.uk/wp-content/uploads/sites/93/2019/09/CyberTroop-Report19.pdf	Worldwide, it is most common for state-sponsored actors to focus on influencing domestic views through social media influence campaigns. However, at least seven have made influence efforts abroad and have well-developed capabilities, and at least a dozen others likely have well-developed capability to do so. If a capable external state actor does have an interest in suppressing or distorting information deemed relevant, assess the state actors' capabilities on relevant platforms in relevant languages and assess how that might distort results from this method. Although the method has several features that make it resistant to distortion, tamper features will be less effective if the state actor is skilled at producing viral content or has the capability to saturate a social media discourse space with disinformation. If a capable state actor conducts influence operations in the country of interest, consider conducting testing or retrospective analysis to determine if informative social media is likely to be or remain published.

Platform Policy

Social media–based methods are most likely to be effective when usage policy for suitable social media platforms does not prohibit an organization from employing this methodology, nor does it limit data transfer in ways that render it ineffective. Table B.3 discusses these factors in detail.

Table B.3
Platform Policy

Platform Policy	Notes
Platform does not prohibit an organization from accessing post data and does not prohibit using post data to scan for developing public events. Twitter, "Developer: Developer Agreement and Policy," webpage, March 10, 2020, https://developer.twitter.com/en/developer-terms/agreement-and-policy YouTube, "YouTube API Services Terms of Service: Usage and Quotas," webpage, last updated August 27, 2020, https://developers.google.com/youtube/v3/getting-started#quota	Social media platforms often have detailed rules in their terms of service for what entities are allowed to access the platform and for what purpose. Read each platform's terms of service carefully before applying this method to it. As formulated, this method provides a tool for gaining awareness on noteworthy public events as they are happening, which is consistent with the terms of service for most broadcast-oriented platforms. However, an organization may need to apply for approval before applying it.
Platform does not prohibit downloading posts in bulk from strangers. See terms of service citations above.	Some platforms do not permit accessing a user's posts without opt-in permission from that user. Such platforms—including Facebook, among others—are less likely to be suitable for this method because it would require individually recruiting 200,000+ people.
Platform does not prohibit downloading posts fast enough for same-day event monitoring. Twitter, "Developer: Rate Limits, Overview, Standard API Rate Limits Per Window," webpage, undated, https://developer.twitter.com/en/docs/basics/rate-limits "YouTube API Services Terms of Service: Usage and Quotas," webpage, last updated August 27, 2020, https://developers.google.com/youtube/v3/getting-started#quota	This methodology requires a significant volume of data, but social media platform APIs typically limit the rate at which data can be downloaded. Platforms are only suitable for this method if an acceptable amount of data can be downloaded daily.

Analyst Capabilities

Social media–based methods are most likely to be effective when analysts are culturally fluent, technologically supported, and have access to sufficient regional database material to facilitate cross-checking. Table B.4 discusses these factors in detail.

Table B.4
Analyst Capabilities

Analyst Capabilities	Notes
Analysts understand the languages commonly used in this locale and are culturally fluent in the idioms, abbreviations, and references common in the internet dialect common in this region. Find data here: Central Intelligence Agency, "The World Factbook," webpage, undated, https://www.cia.gov/library/publications/the-world-factbook Eberhard, Simons, and Fennig, undated, https://www.ethnologue.com	A minimum prerequisite for this methodology is that analysts have the cultural fluency to understand the languages typically used to compose social media posts in the geographic area of interest. There may be users writing in English who claim to be writing from the region, but such accounts should not be used unless English is a commonly spoken language in the area. Cultural fluency goes beyond academic knowledge of the language. Social media posts are often written in an online dialect that requires additional cultural context and knowledge of idioms. To give three examples: (1) users writing in non-Latin alphabet languages may develop conventions for using Latin alphabets, such as when the Arabic letter Ayn is represented with a "3"; (2) emoji may develop different connotations, such as the offensive meaning ascribed to the "thumbs up" emoji in some parts of West Africa and Europe; and (3) shorthand notations may develop and evolve over time, such as the "555" homophone to signify laughter in Thai or the "JTM" abbreviation for "I love you" in French.
Analysts have the technical capability to pull and analyze bulk data from platforms commonly used in this locale. For patterns of social media usage in specific countries, see We Are Social, "Digital in 2020," webpage, undated, https://wearesocial.com/digital-2020 Website popularity data: Alexa, "The Top 500 Sites on the Web," webpage, undated, https://www.alexa.com/topsites/countries	Analysts must have information technology, technical support, and social media platform access sufficient to download, process, and conduct an automated scan of 8+ gigabytes of social media textual every day in 16 hours or less and provide analysts with multiplatform access during the remaining eight hours. Social media platform popularity varies by geographic area. This method will be most successful when analysts have technical capability to pull data from platforms that are among the most popular in a given area.

Table B.4—Continued

Analyst Capabilities	Notes
Geographic databases have enough coverage of this region to enable location fact-checking. Find data here: Google, "Google Maps," undated, https://www.google.com/maps Find data here: Flickr, website, undated, https://www.flickr.com	Misinformation and disinformation are serious concerns with social media data. Fact-checking information using geographic databases is a powerful method for lowering these risks, because geography is detail-rich and relatively stable over time. However, location-based databases have limited coverage in some regions. Assuming that satellite-based databases cannot fill in the gap, the method may be less tamper-resistant in areas where databases are lacking.

APPENDIX C

Estimating Analyst Level of Effort Requirements

In Chapter Three, we discussed how we estimate analyst level of effort requirements. This appendix discusses how we derived that estimate in more detail.

In conducting manual analysis of social media, we were consistently able to analyze each post and its linked media in about ten minutes or less. Table C.1 reports the distribution of time needed to analyze posts that did and did not contain relevant information.

Based on these figures, we estimated the potential manpower requirements for the social media analysis method. For our figures, we assumed that each employee would be able to devote seven hours per workday to analyzing posts, with the remaining hour devoted to other job tasks and breaks.

For perspective, let us start by imagining that Step 3—conduct automated daily scans for elevated keyword use—was dropped from the method entirely, such that analysis was conducted manually for every post that our 200,000-user sample wrote on any given day

Table C.1
Time to Complete Manual Analysis of Posts and Linked Media

	Time in Minutes		
	10th Percentile	50th Percentile	90th Percentile
Post and linked media had no relevant information	3	4	8
Post and/or linked media held relevant information	5	6	10

during the 2017–2018 period. We estimate that 70–230 STE would be necessary to analyze every post in our sample. However, this only accounts for one language. If similar samples were drawn for the top 20 languages that represent 64 percent of the world's population, the STE requirement could rise proportionally to 1,400–4,600 STE. If similar samples were drawn for each of the 200 languages that represent 88 percent of the world's population, the STE requirement could rise to 14,000–46,000.

Now imagine that analysts were only asked to read posts that contained keywords. In our case study sample, we estimate that, for 95 percent of all days, it would have taken 1–29 STE to read every post that contained a keyword. Assuming that samples in all other languages behaved like our test case, that translated to 20–580 STE to read every keyword-bearing post in the 20 most spoken languages and 200–5,800 STE to read every keyword-bearing post in the 200 most spoken languages.

However, Step 3 in our method also has a procedure for determining which days are more likely to have had a chemical weapons incident. Imagine that staff could be allocated to social media analysis on an as-needed basis. Perhaps these staff conducted a different kind of analysis that requires similar skills and that these staff can be rapidly re-tasked to our social media procedure if the algorithm detects elevated keyword use. Peak STE usage would be the same as above. However, total annual STE usage would be much lower. Assuming this kind of rapid re-tasking, a sample that is like our test case data, and a 230-day work year, our social media protocol could only require the cost equivalent of 3–5 STE for one language. Assuming that samples in all other languages behaved like our test case, that amounts to the cost equivalent of 60–100 STE to apply the method to the 20 most spoken languages and 600–1,000 STE to apply the method to the 200 most spoken languages.

These calculations do not include the other staff needed to support analysts, including administrative and information technology staff.

APPENDIX D

Protecting Analysts from Secondary Trauma

Material presented in this appendix is adapted from guidance developed as a resource for RAND research staff conducting research on topics that present risks of secondary trauma.

Introduction and Purpose

The mental health field has long focused on assessing the impact of traumatic events on individuals who experience them, and, more recently, the focus has expanded somewhat to include those indirectly affected by trauma. In particular, the concepts of "vicarious traumatization" or "secondary traumatic stress" were developed and studied to capture the impact of indirect trauma exposure on such professionals as mental health practitioners, educators, journalists, researchers, and human rights activists.

In the course of their work, policy researchers can encounter potential exposure to secondary trauma. For example, national security–related research requiring data collection on exposure to combat trauma, wartime trauma and atrocities, terrorism, or natural disasters can present risks to research staff. Similarly, studies that involve interviews and discussions with survivors of loss (suicide or mass violence), victims of interpersonal violence or assault, persons who have experienced sexual assault or exploitation, and caregivers and providers for individuals who have been catastrophically wounded can also present risks. Whether it is gathering data on the circumstances of these events or asking individuals exposed to them to share their stories of grief, loss, and recovery, these forms of elicitation can expose researchers to the trauma. However, to date, there has been no guidance on how researchers and supervisors can address these exposures. This appendix was developed to acknowledge this job-related stress and provide information about the types of problems commonly experienced and recommended strategies for mitigating them. The intent is to equip research staff with tools and resources for identifying symptoms and practicing self-care and to provide guidance to supervisors, including project leaders, on approaches to structuring work to best support the research team.

Defining Traumatic Content

Psychologists and psychiatrists typically define a trauma within the context of diagnosing posttraumatic stress disorder. This definition distinguishes trauma from everyday stressors by specifying a trauma to be "actual or threatened death, serious injury, or sexual violence." The definition acknowledges that exposures can be of various types: direct personal exposures, witnessing (in person) trauma to others, learning that a family member or close friend was exposed to a trauma, or indirect repeated or extreme exposure to aversive details of a traumatic event. This last type of exposure is most relevant to professionals who encounter trauma in their work lives, and these work exposures are in fact specifically called out as applicable in the *Diagnostic and Statistical Manual of Mental Disorders, 5th ed.* (DSM-5) descriptions.

Types of Reactions

Several different terms used to describe professionals' reactions to traumatic work content have evolved over time: vicarious trauma, secondary traumatic stress, and compassion fatigue. We will use the term *secondary traumatic stress* because the terms overlap, and there is currently no consensus on terminology. In addition, the more general concept of professional burnout is pertinent to trauma exposure at work.

Secondary traumatic stress is generally thought of as the symptoms of posttraumatic stress disorder. These include four clusters of related symptoms:

- intrusive symptoms (nightmares, unwanted upsetting thoughts, flashbacks, feeling upset when reminded of the trauma)
- avoidance (wanting to push away trauma-related thoughts or feelings or avoid external reminders of the trauma)
- negative thoughts and feelings (distorted beliefs about self or others, ongoing strong negative emotions, less interest in activities, feeling detached from others)
- arousal and reactivity (more irritability, risky behavior, hypervigilance, startle reaction, trouble sleeping or concentrating since the trauma occurred).

Research on therapeutic professionals show the following risk factors for secondary traumatic stress, which could likely be applicable to researchers:

- amount, frequency, and proportion of work involving exposure to trauma content
- personal trauma history
- low level of work support
- low level of social support.

Professional burnout is another potential consequence of work-related trauma exposure, although it seems to be more related to general work stress (unfair treatment, unmanageable workload, lack of role clarity, unreasonable time pressure, and poor communication) than to trauma content per se. Burnout does not involve the specific symptoms of posttraumatic stress disorder, but rather general feelings of hopelessness and being ineffectual at work, gradually leading to a state of emotional exhaustion that can lead to employment turnover.

Assessment Tools

Several different assessment tools exist for research and self-assessment purposes for these overlapping concepts. The most comprehensive is likely the Professional Quality of Life Scale, which includes subscales for secondary traumatic stress and burnout, as well as a positive scale assessing work satisfaction. Although it is geared toward helping professionals, such as social workers, the concepts assessed may be relevant to policy researchers as well. Items tapping secondary traumatic stress include general items ("I jump or am startled by unexpected sounds") and items specific to trauma work ("I think I might be affected by the traumatic stress of [those I help]," "As a result of my [work], I have intrusive, frightening thoughts"). Examples of burnout items include "I feel trapped by my job [as a helper]," or "I feel worn out because of my work [as a helper]."

What Researchers Can Do to Take Care of Themselves

The most common strategies to mitigate secondary traumatic stress are basic self-care strategies. One resource is the Taking Care of Yourself fact sheet, which describes three categories of strategies: awareness, balance, and connection. It suggests daily self-care ideas, such as practicing brief relaxation strategies, and keeping a journal, practicing good sleep routines, and work-related strategies, such as checking in with colleagues, maintaining boundaries, and saying no.

Some specific suggestions include:

- Practice good self-care.
- Take control over one's workflow, such as by choosing the timing, location, and modality of the work-related trauma exposures. Different approaches work for different people, but some possibilities include:
  - Consider working on trauma content in the mornings, with time to change your mindset before going home.
  - Consider limiting work in evenings or when at home.

Example: A researcher and mother to young children was working on a project that involved a review of the empirical literature on suicide among young children. She found the project work to be difficult emotionally—she found herself thinking a lot about what it would be like if one of her own children died by suicide. After working a few hours on the project, she would feel sad and anxious. She found that she needed to do this work for more limited amounts of time, earlier in the day, well before picking her children up from school. She let the project leader know that she was going to be dividing up her time this way so that they would understand why she would not be working on the child suicide project in the afternoons.

- Consider spacing out the trauma-related work rather than massing it into a few days.
- Consider building in self-care breaks into your daily work schedule or developing a detailed self-care plan.
- Consider debriefing with a coworker or supervisor after extended periods of time working on this subject matter.

- Discuss expectations on your workflow or timeline with your supervisor, especially if your supervisor seems to be unreasonable. Consider talking to the supervisor about the impact of the work on your mood.
- Seek social support (e.g., interact with coworkers, friends, or family).
- Maintain solid work-home boundaries, trying to protect your personal time.
- If the work involves travel, try to plan the trip to allow for adequate down time for sleep and self-care activities.
- If you need additional support, consider seeking professional help.

Example: A researcher was working on a project that involved exposure to social media content, including video, audio, and text of the effects of chemical weapons employment on victims in Syria. She found the depictions of violence very disturbing and took steps to exert control over her exposure. For example, she separated exposure to audio from video, as she found them more upsetting together, especially for native speakers (and she limited exposure to what was absolutely required). She also previewed imagery in videos and turned off autoplay so she could prepare for, or work around, images to the extent possible before exposure.

What Supervisors Can Do to Support Their Research Teams

A resource developed for clinical supervisors contains some ideas for supervisor competencies, including the following which may be relevant a research supervisor role.

Recognize the signs of secondary traumatic stress.

- Address observed signs in a supportive manner (normalizing responses, rather than demeaning them).
- Delineate support options available from the organization and other resources.

In discussions with project teams based on their experiences with this type of work, and building on the ideas above, we suggest the following for supervisors:

- Be sure to explain the work and its traumatic content when selecting and onboarding new project team members. Some staff members may self-select out of the project if they do not want this kind of exposure at work. At the same time, do not make assumptions about whether staff will want to do certain types of work. Lay it out clearly, and let them make their own decision.

- Understand that different people have different sensitivities. One person might not react at all to a certain kind of trauma, whereas another person might find it very difficult. Do not assume that something that does not bother you will not bother other people.

Regularly acknowledge that the work can be stressful; check in with staff often.

- Consider workflow and pace, and ensure the staff members have some control over these in relation to the traumatic content so they can modulate their exposures.
- If the work involves travel, encourage a travel schedule that allows for adequate time for sleep, exercise, and other self-care activities.
- Consider periodic voluntary formal and informal check-ins with staff that might include:

  - allowing expression of stress related to work content
  - providing a nonjudgmental, nonhierarchical, no-consequences atmosphere

Example: Survey researchers develop project-specific plans when hiring and training field staff on projects involving difficult populations or upsetting topics. They begin with training modules that try to prepare staff for the types of things they might see or hear on the project. During the field work, they try to balance the workload across staff so no one person is overburdened by the more difficult content. Most projects bring together the field staff for regular weekly or biweekly meetings, where part of the agenda is to share stories and experiences from the field, encouraging staff to talk to one another and to their supervisor. Key elements are orienting staff to the idea that things they might see and hear can be very upsetting, having open discussion of self-care, and clearly conveying the message that there will be no penalties if they need a break from a certain project or site.

- allowing for peer-to-peer support (with or without you present)
- discussing researcher self-care plans
- possibly scheduling occasional informal get togethers to offer opportunities for nonwork-related support and social time.

Resources

Two excellent resources explain the material in this appendix in clear and accessible language. These may be appropriate for use in onboarding staff members to a project that will include exposure to traumatic content.

An article from Amnesty International describes how secondary trauma can affect activists and reporters.[1] The article has a more general and less mental-health focus and therefore might be more accessible for some.

The National Child Traumatic Stress Network has developed a landing page and resources related to secondary traumatic stress.[2] Resources include a general introduction, several fact sheets for different types of professionals and contexts, a few webinars on the impact on professionals, and a "taking care of yourself" fact sheet. Many of these are more oriented toward clinicians (but still contain some useful material), but some are more general.

---

[1] Rossalyn Warren, "The Hidden Victims of Repression—How Activists and Reporters Can Protect Themselves from Secondary Trauma," Amnesty International website, February 20, 2019.

[2] National Child Traumatic Stress Network, "Secondary Traumatic Stress," webpage, undated.

APPENDIX E

Social Media Event-Detection Literature Review

Background

Big data is the umbrella term for changes in the modern data environment that challenge traditional analytic approaches and present new opportunities for data-driven operations. The 5V definition of big data outlines five characteristics of big data:[1]

- **Volume:** Data are no longer scarce. The challenge is sorting through a wealth of data to derive high-value information, rather than collecting useful data.
- **Velocity:** Data are generated in real time and can change quickly. The challenge is analyzing data quickly enough to be relevant and determining how best to chunk a continuous stream of data for analysis.
- **Variety:** Data consist of any combination of textual, visual, and audio data varied combinations from varied sources. The challenge is developing a common representation of data, not collecting data designed to fit the format of a tabular data set.
- **Value:** Data may not have been intended to be data and may require significant processing to extract high-value information

[1] Gema Bello-Orgaz, Jason J. Jung, and David Camacho, "Social Big Data: Recent Achievements and New Challenges," *Information Fusion*, Vol. 28, March 2016; Yuri Demchenko, Canh Ngo, and Peter Membrey, "Architecture Framework and Components for the Big Data Ecosystem," Draft Version 0.2, September 12, 2013.

from them. The challenge is recognizing the information that can be extracted and building refinement processes to extract it.
- **Veracity:** Data may include portions that are incomplete, inaccurate, or even intended to deceive. The challenge is building methods that are robust to imperfections and able to assemble accurate information from contradictory pieces.

Social media is a promising application of the big-data paradigm.[2] Smartphones have placed sophisticated sensing devices in pockets around the world (volume, variety), and social media platforms provide people with the capability to broadcast thoughts and experiences in real time (volume, velocity). Social media is not purpose-built to provide information for data analysis, but information can be extracted from it with proper refining and vetting processes (value, veracity).

This appendix overviews patterns in how researchers have applied social media–analysis techniques to the task of detecting, verifying, and gathering information on emerging events.

How We Built the Appendix

To develop this appendix, we queried Web of Knowledge and Google Scholar for relevant articles, using four rounds of queries. First, we combined the search phrase "social media" with such search phrases as "developing event," "emerging event," "event detection," or "security OR defense." Second, we conducted a second round of queries for highly cited articles using the search phrase "social media" and then manually scanned through them to determine if our first round of search generally caught the articles likely relevant to our study. Third, we queried RAND's report database using the search phrase "social media" and manually scanned through the results to find pieces consistent with the documents found during the first two rounds.

[2] Diogo Nolasco and Jonice Oliveira, "Subevents Detection Through Topic Modeling in Social Media Posts," *Future Generation Computer Systems*, Vol. 93, April 2019.

Fourth, we conducted ad hoc queries while drafting the text to confirm specific statements.

Our literature review suggests at least five topic areas in which emerging event detection, vetting, and intelligence has become a focus of study: (1) marketing, (2) public health, (3) disaster intelligence, (4) public order and political influence, and (5) misinformation spread.

Use of Social Media Emerging Event Analysis Across Five Topic Areas

This section discusses the use of social media–emerging event analysis across five topic areas. First, *marketing practitioners* strive to channel social media activity in ways that promote brands and link online activity of desired offline outcomes. Second, *public health practitioners* strive to educate the public and detect emerging health issues so remedial action may be taken. Third, *disaster recovery practitioners* strive to establish situational awareness under circumstances where traditional information flow has been disrupted. Fourth, *public order and political influence practitioners* strive to gain situational awareness on emerging public safety incidents and social media activities that may lead to public safety incidents. They may also seek to either influence populations or prevent influence. Finally, *misinformation researchers* strive to understand the spread of misinformation and its consequences for polarization.

Marketing

The marketing industry routinely uses social media as part of standard operating practices. Social media has become a crucial channel for word-of-mouth advertising,[3] and review websites have become a focal

[3] Ana Babić Rosario, Francesca Sotgiu, Kristine De Valck, and Tammo Bijmolt, "The Effect of Electronic Word of Mouth on Sales: A Meta-Analytic Review of Platform, Product, and Metric Factors," *Journal of Marketing Research*, Vol. 53, No. 3, June 2016.

point for product-experience monitoring.[4] New fields of marketing have developed to reach consumers through search engines, viral advertising content that consumers will spontaneously distribute to one another, and social media to provide low-latency market intelligence.[5] Marketing has perhaps gone the furthest in matching online behavior to offline behaviors, developing techniques to link social media data to customer profitability,[6] recruitment,[7] and other marketing outcomes.

Public Health

In addition to documenting the direct impacts of social media on mental health,[8] health care providers have explored how social media analysis can fit into a public health strategy. Recognizing that patients often turn to the internet for health information, a significant body of research has explored how to use social media for public health

---

4 Zheng Xiang, Qianzhou Du, Yufeng Ma, and Weiguo Fan, "A Comparative Analysis of Major Online Review Platforms: Implications for Social Media Analytics in Hospitality and Tourism," *Tourism Management*, Vol. 58, February 2017.

5 Cait Lamberton and Andrew T. Stephen, "A Thematic Exploration of Digital, Social Media, and Mobile Marketing: Research Evolution from 2000 to 2015 and an Agenda for Future Inquiry," *Journal of Marketing*, Vol. 80, No. 6, November 2016; Ming Ni, Qing He, and Jing Gao, "Forecasting the Subway Passenger Flow Under Event Occurrences with Social Media," *IEEE Transactions on Intelligent Transportation Systems*, Vol. 18, No. 6, June 2017.

6 Ashish Kumar, Ram Bezawada, Rishika Rishika, Ramkumar Janakiraman, and P. K. Kannan, "From Social to Sale: The Effects of Firm-Generated Content in Social Media on Customer Behavior," *Journal of Marketing*, Vol. 80, No. 1, January 2016.

7 Wenger et al., 2019.

8 John A. Naslund, Kelly A. Aschbrenner, Lisa A. Marsch, and Stephen J. Bartels, "The Future of Mental Health Care: Peer-to-Peer Support and Social Media," *Epidemiology and Psychiatric Sciences*, Vol. 25, No. 2, April 2016; Jean M. Twenge, Thomas E. Joiner, Megan L. Rogers, and Gabrielle N. Martin, "Increases in Depressive Symptoms, Suicide-Related Outcomes, and Suicide Rates Among U.S. Adolescents After 2010 and Links to Increased New Media Screen Time," *Clinical Psychological Science*, Vol. 6, No. 1, 2018; Heather Cleland Woods and Holly Scott, "#Sleepyteens: Social Media Use in Adolescence Is Associated with Poor Sleep Quality, Anxiety, Depression and Low Self-Esteem," *Journal of Adolescence*, Vol. 51, August 2016.

education and efforts to counter misinformation.[9] Researchers have also experimented with social methods for detecting emerging infectious disease and medical product safety concerns.[10] However, this is a difficult task because consumers generally do not have the technical training to conceptualize their experiences in ways that will yield useful public health data.

Disaster Intelligence

During disasters, routine channels of information flow are often disrupted, and ground conditions change rapidly. The flexibility, decentralization, and mobility of social media show promise under such conditions, potentially providing a robust option for regaining

---

[9] Rebecca L. Collins, Eunice C. Wong, Joshua Breslau, M. Audrey Burnam, Matthew Cefalu, and Elizabeth Roth, "Social Marketing of Mental Health Treatment: California's Mental Illness Stigma Reduction Campaign," *American Journal of Public Health*, Vol. 109, Supplement 3, June 2019; Kathryn H. Jacobsen, A. Alonso Aguirre, Charles L. Bailey, Ancha V. Baranova, Andrew T. Crooks, Arie Croitoru, Paul L. Delamater, Jhumka Gupta, Kylene Kehn-Hall, Aarthi Narayanan, Mariaelena Pierobon, Katherine E. Rowan, J. Reid Schwebach, Padmanabhan Seshaiyer, Dann M. Sklarew, Anthony Stefanidis, and Peggy Agouris, "Lessons from the Ebola Outbreak: Action Items for Emerging Infectious Disease Preparedness and Response," *Ecohealth*, Vol. 13, No. 1, March 2016; Zhen Wang, Chris T. Bauch, Samit Bhattacharyya, Alberto d'Onofrio, Piero Manfredi, Matjaz Perc, Nicola Perra, Marcel Salathé, and Dawei Zhao, "Statistical Physics of Vaccination," *Physics Reports*, Vol. 664, November 2016.

[10] Carrie E. Pierce, Khaled Bouri, Carol Pamer, Scott Proestel, Harold W. Rodriguez, Hoa Van Le, Clark C. Freifeld, John S. Brownstein, Mark Walderhaug, I. Ralph Edwards, and Nabarun Dasgupta, "Evaluation of Facebook and Twitter Monitoring to Detect Safety Signals for Medical Products: An Analysis of Recent FDA Safety Alerts," *Drug Safety*, Vol. 40, No. 4, 2017; Ovidiu Şerban, Nicholas Thapen, Brendan Maginnis, Chris Hankin, and Virginia Foot, "Real-Time Processing of Social Media with SENTINEL: A Syndromic Surveillance System Incorporating Deep Learning for Health Classification," *Information Processing and Management*, Vol. 56, No. 3, May 2019; John van Stekelenborg, John, Johan Ellenius, Simon Maskell, Tomas Bergvall, Ola Caster, Nabarun Dasgupta, Juergen Dietrich, Sara Gama, Davis Lewis, Victoria Newbould, Sabine Brosch, Carrie E. Pierce, Gregory Powell, Alicia Ptaszyńska-Neophytou, Antoni F. Z. Wiśniewski, Phil Tregunno, G. Niklas Norén, and Munir Pirmohamed, "Recommendations for the Use of Social Media in Pharmacovigilance: Lessons from IMI WEB-RADR," *Drug Safety*, Vol. 42, No. 12, December 2019.

situation awareness.[11] Most directly, social media can provide an auxiliary communication pathway to call for help,[12] especially when traditional response communication infrastructure is disrupted.[13] Researchers have also explored approaches for using social media to gain rapid awareness of infrastructure failures and map disruption to daily life,[14] which may provide information faster than traditional in-person professional assessment. Ground truth can be difficult to establish during disasters. Some work has built credibility-assessment methods to address this limitation of disaster social methods.[15] Other methods work in tandem with remote sensing, finding priority locations where limited satellite, drone, and lidar resources can be most effectively used.[16]

Public Order and Political Influence

Social media has become an important front in both the incitement and prevention of hate crime specifically,[17] as well as in predictive

---

[11] Jingchao Yang, Manzhu Yu, Han Qin, Mingyu Lu, and Chaowei Yang, "A Twitter Data Credibility Framework—Hurricane Harvey as a Use Case," *ISPRS International Journal of Geo-Information*, Vol. 8, No. 3, 2019; Yeung et al., 2020.

[12] Jyoti Prakash Singh, Yogesh K. Dwivedi, Nripendra P. Rana, Abhinav Kumar, and Kawaljeet Kaur Kapoor, "Event Classification and Location Prediction from Tweets During Disasters," *Annals of Operations Research*, Vol. 283, No. 1, 2017.

[13] Jacey Fortin, "When Disaster Hits and Landlines Fail, Social Media Is a Lifeline," *New York Times*, 2017.

[14] Fan and Mostafavi, 2019; Marcelo Mendoza, Bárbara Poblete, and Ignacio Valderrama, "Nowcasting Earthquake Damages with Twitter," *EPJ Data Science*, Vol. 8, No. 3, 2019, p. 3.

[15] Jingchao Yang et al., 2019.

[16] Kashif Ahmad, Konstantin Pogorelov, Michael Riegler, Nicola Conci, and Pål Halvorsen, "Social Media and Satellites," *Multimedia Tools and Applications*, Vol. 78, No. 3, February 2019; Julian F. Rosser, Didier G. Leibovici, and Margaret J. Jackson, "Rapid Flood Inundation Mapping Using Social Media, Remote Sensing and Topographic Data," Natural Hazards, Vol. 87, No. 1, 2017.

[17] Matthew L. Williams and Pete Burnap, "Cyberhate on Social Media in the Aftermath of Woolwich: A Case Study in Computational Criminology and Big Data," *British Journal of Criminology*, Vol. 56, No. 2, March 2016; Matthew L. Williams, Pete Burnap, Amir Javed, Han Liu, and Sefa Ozalp, "Hate in the Machine: Anti-Black and Anti-Muslim Social Media

policing more generally.[18] Public security practitioners have also begun building systems that use social media to scan for emerging public order events,[19] and they have been used to direct staff toward the security camera feeds most likely to provide visibility on that event.[20] Such systems may also strive to aggregate social media and/or news media to provide more complete situational awareness.[21] Going beyond reactive social media postures, some state actors have built sophisticated apparatuses for influencing social media to achieve desired real-world outcomes, such as encouraging pro-government views, suppressing

---

Posts as Predictors of Offline Racially and Religiously Aggravated Crime," *British Journal of Criminology*, Vol. 60, No. 1, January 2020.

[18] Lara Vomfell, Wolfgang Karl Härdle, and Stefan Lessmann, "Improving Crime Count Forecasts Using Twitter and Taxi Data," *Decision Support Systems*, Vol. 113, September 2018; Dingqi Yang, Terence Heaney, Alberto Tonon, Leye Wang, and Philippe Cudré-Mauroux, "CrimeTelescope: Crime Hotspot Prediction Based on Urban and Social Media Data Fusion," *World Wide Web*, Vol. 21, 2018.

[19] Chao Zhang, Guangyu Zhou, Quan Yuan, Honglei Zhuang, Yu Zheng, Lance Kaplan, Shaowen Wang, and Jiawei Han, "Geoburst: Real-Time Local Event Detection in Geo-Tagged Tweet Streams," in *SIGIR '16: Proceedings of the 39th International ACM SIGIR Conference on Research and Development in Information Retrieval*, July 2016.

[20] Zheng Xu, Hui Zhang, Chuanping Hu, Lin Mei, Junyu Xuan, Kim-Kwang Raymond Choo, Vjayan Sugumaran, and Yiwei Zhu, "Building Knowledge Base of Urban Emergency Events Based on Crowdsourcing of Social Media," *Concurrency and Computation: Practice and Experience*, Vol. 28, No. 15, October 2016; Zheng Xu, Lin Mei, Zhihan Lv, Chuaping Hu, Xiangfeng Luo, Hui Zhang, and Yunhuai Liu, "Multi-Modal Description of Public Safety Events Using Surveillance and Social Media," *IEEE Transactions on Big Data*, Vol. 5, No. 4, October–December 2019; Zheng Xu, Yunhuai Liu, Junyu Xuan, Haiyan Chen, and Lin Mei, "Crowdsourcing Based Social Media Data Analysis of Urban Emergency Events," *Multimedia Tools and Applications*, Vol. 76, No. 9, 2017.

[21] Ian McCulloh and Kathleen M. Carley, "Detecting Change in Longitudinal Social Networks," *Journal of Social Structure*, Vol. 12, No. 1, 2011; Matthew Schehl, "NPS Team Turns to Machine Learning to Predict Social Unrest," Naval Postgraduate School webpage, January 16, 2019; Xueming Qian, Mingdi Li, Yayun Ren, and Shuhui Jiang, "Social Media Based Event Summarization by User–Text–Image Co-Clustering," *Knowledge-Based Systems*, Vol. 164, No. 15, January 2019; Zhenguo Yang, Qing Li, Wenyin Liu, and Jianming Lv, "Shared Multi-View Data Representation for Multi-Domain Event Detection," *IEEE Transactions on Pattern Analysis and Machine Intelligence*, Vol. 42, No. 5, January 2019.

dissent, or encouraging negative sentiment toward adversaries.[22] Worldwide, these cyberpropaganda activities are most often deployed in a domestic context. However, some states deploy these techniques against populations in other countries, striving to weaken rival states, improve standing in foreign relations, or support physical warfare activities.[23] In turn, this has generated a body of research on countering influence campaigns,[24] especially in terms of detecting, understanding, and fighting automated bot tactics.[25]

Diffusion of Misinformation

As just discussed, social media's impact on public knowledge and worldview is a concern across multiple use cases. This concern has produced a body of scholarship focused on the information dynamics of misinformation. Some work focuses on how the behavioral dynamics of social media encourage misinformation, such as the role

---

[22] Adam Badawy, Aseel Addawood, Kristina Lerman, and Emilio Ferrara, "Characterizing the 2016 Russian IRA Influence Campaign," *Social Network Analysis and Mining*, Vol. 9, No. 1, 2019, article 31; Samantha Bradshaw and Philip N. Howard, *The Global Disinformation Order: 2019 Global Inventory of Organised Social Media Manipulation*, Oxford: Computational Propaganda Project at the Oxford Internet Institute, 2019; Martin Kragh and Sebastian Åsberg, "Russia's Strategy for Influence Through Public Diplomacy and Active Measures: The Swedish Case," *Journal of Strategic Studies*, Vol. 40, No. 6, 2017; William Marcellino, Meagan L. Smith, Christopher Paul, and Lauren Skrabala, *Monitoring Social Media: Lessons for Future Department of Defense Social Media Analysis in Support of Information Operations*, Santa Monica, Calif.: RAND Corporation, 2017, RR-1742-OSD, 2017; T. Camber Warren, "Not by the Sword Alone: Soft Power, Mass Media, and the Production of State Sovereignty," *International Organization*, Vol. 68, No. 1, January 2014.

[23] Bradshaw and Howard, 2019; Adam Entous, Ellen Nakashima, and Greg Jaffe, "Kremlin Trolls Burned Across the Internet as Washington Debated Options," *Washington Post*, December 25, 2017.

[24] Elizabeth Bodine-Baron, Todd C. Helmus, Andrew Radin, and Elina Treyger, *Countering Russian Social Media Influence*, Santa Monica, Calif.: RAND Corporation, RR-2740-RC, 2018; Helmus et al., 2018; Marcellino et al., 2020.

[25] Emilio Ferrara, Onur Varol, Clayton Davis, Filippo Menczer, and Alessandro Flammini, "The Rise of Social Bots," *Communications of the ACM*, Vol. 59, No. 7, July 2016; Bjarke Mønsted, Piotr Sapieżyński, Emilio Ferrara, and Sune Lehmann, "Evidence of Complex Contagion of Information in Social Media: An Experiment Using Twitter Bots," *PloS One*, Vol. 12, No. 9, September 2017.

of confirmation bias and tendency of like-minded people to form social cliques.[26] Other work tracks the characteristics and life cycles of rumors,[27] often focusing on strategies for measuring authenticity and credibility.[28]

Common Approaches

Across different topic areas, researchers typically apply some combination of different data-analysis techniques. This section discusses six techniques that appear in various combinations in the pieces cited earlier in this report:[29] (1) place-name surge detection,

---

[26] Michela Del Vicario, Alessandro Bessi, Fabiana Zollo, Fabio Petroni, Antonio Scala, Guido Caldarelli, H. Eugene Stanley, and Walter Quattrociocchi, "The Spreading of Misinformation Online," *Proceedings of the National Academy of Sciences of the United States of America*, Vol. 113, No. 3, January 2016; Wang et al., 2016; Stephan Winter, Miriam J. Metzger, and Andrew J. Flanagin, "Selective Use of News Cues: A Multiple-Motive Perspective on Information Selection in Social Media Environments," *Journal of Communication*, Vol. 66, No. 4, August 2016.

[27] Sarah A. Alkhodair, Steven H. H. Ding, Benjamin C. M. Fung, and Junqiang Liu, "Detecting Breaking News Rumors of Emerging Topics in Social Media," *Information Processing and Management*, Vol. 57, No. 2, March 2020; Svitlana Volkova, Kyle Shaffer, Jin Yea Jang, and Nathan Hodas, "Separating Facts from Fiction: Linguistic Models to Classify Suspicious and Trusted News Posts on Twitter," in Regina Barzilay and Min-Yen Kan, eds., *Proceedings of the 55th Annual Meeting of the Association for Computational Linguistics* (Vol. 2: *Short Papers)*, July 2017; Arkaitz Zubiaga, Maria Liakata, Rob Procter, Geraldine Wong Sak Hoi, and Peter Tolmie, "Analysing How People Orient To and Spread Rumours in Social Media by Looking at Conversational Threads," *PloS One*, Vol. 11, No. 3, 2016; Arkaitz Zubiaga, Maria Liakata, and Rob Procter, "Exploiting Context for Rumour Detection in Social Media," *Social Informatics*, 2017; Arkaitz Zubiaga, Ahmet Aker, Kalina Bontcheva, Maria Liakata, and Rob Procter, "Detection and Resolution of Rumours in Social Media: A Survey," *ACM Computing Surveys*, Vol. 51, No. 2, February 2018.

[28] Dinesh Kumar Vishwakarma, Deepika Varshney, and Ashima Yadav, "Detection and Veracity Analysis of Fake News via Scrapping and Authenticating the Web Search," *Cognitive Systems Research*, Vol. 58, December 2019; Zubiaga et al., 2017; Xichen Zhang and Ali A. Ghorbani, "An Overview of Online Fake News: Characterization, Detection, and Discussion," *Information Processing and Management*, Vol. 57, No. 2, March 2020.

[29] In addition to works cited in the previous section, methodology focused pieces informing this section include the following: Choi and Park, 2019; Mahmud Hasan, Mehmet A. Orgun, and Rolf Schwitter, "Real-Time Event Detection from the Twitter Data Stream Using the

(2) activity keyword surge detection, (3) machine learning from labeled training sets, (4) machine learning from affinity clustering, (5) social verification, and (6) blended computer-human systems. These techniques are often used in tandem, creating a diverse array of technical approaches. Place-name surge detection techniques search for abrupt increases in the use of proper names for a specific geographic location. Activity surge detection techniques search for abrupt increases in the use of words that describe a specific kind of event. Deductive machine-learning techniques provide an algorithm with human-categorized examples of an event and then has the algorithm search for new events that it sees as similar to the examples of a specific category of events. Inductive machine-learning techniques infer categories by clustering posts according to their similarities and then determine the noteworthiness of emerging events based on how the new events fit within the categorization system. Social-verification techniques use interactions between users to determine the post credibility. Blended computer-human intelligence techniques apply computational filtering to find a manageable subset of posts that call for manual examination.

Place-Name Surge Detection

Place-name surge detection scans for sudden increases in words that denote specific geographic locations. Although most social media are not directly geotagged, it is common for people to mention a location when discussing an emerging event. When a particular place name begins appearing in posts at statistically elevated levels, there is a good chance that something noteworthy is happening at that location. This approach works best for place names that are uncommon. For example,

---

TwitterNews+ Framework," *Information Processing and Management*, Vol. 56, No. 3, May 2019; Faria Nazir, Mustansar Ali Ghazanfar, Muazzam Maqsood, Farhan Aadil, Seungmin Rho, and Irfan Mehmood, "Social Media Signal Detection Using Tweets Volume, Hashtag, and Sentiment Analysis," *Multimedia Tools and Applications*, Vol. 78, No. 3, February 2019; Zafar Saeed, Rabeeh Ayaz Abbasi, Onaiza Maqbool, Abida Sadaf, Imran Razzak, Ali Daud, Naif Radi Aljohani, and Guandong Xu, "What's Happening Around the World? A Survey and Framework on Event Detection Techniques on Twitter," *Journal of Grid Computing*, Vol. 17, No. 2, June 2019; Zhicheng Shi and Lilian S. C. Pun-Cheng, "Spatiotemporal Data Clustering: A Survey of Methods," *ISPRS International Journal of Geo-Information*, Vol. 8, No. 3, 2019.

this approach would have difficulty distinguishing between the many cities called Springfield that can be found in Australia, Canada, the UK, the United States, and elsewhere.

Activity Keyword Surge Detection

This scans for sudden increases in words that denote specific kinds of events. It works best if words that describe the event are those that the average social media user would employ and do not have other, more common meanings. Medical terminology is an example of good event descriptors that are not common vocabulary. Charge (electrical or price?), novel (new or a book?), and patient (personal description or person seeking medical care?) are all examples of keywords with multiple common meanings.

Deductive Machine Learning from Labeled Training Set

This exposes a machine-learning model to many examples of posts, relying on human judgment to pre-classify the posts as either having or not having a desired attribute. From these examples, the model learns decision rules for using the features of new posts to guess how they would have been classified. In a sense, this approach is an evolution of the place name and activity keyword scanning techniques just described, except that the algorithm decides which keywords are valuable indicators. This method works best when many examples of pre-classified posts are available and when the machine-learning model is unlikely to start encountering features in new posts that it has never seen before. For example, a social media machine-learning model trained to scan for sports content during football season may have some difficulty correctly classifying posts during the March Madness college basketball playoffs.

Inductive Machine Learning from Cluster Analysis

This technique exposes a machine-learning model to many posts plus strategies for measuring affinity between them. From affinity between posts, the model identifies clusters of posts that all have strong affinity for each other and builds decision rules for classifying posts in a way that corresponds to the clusters observed. Because labeling is inductive, this machine-learning approach is particularly capable of coping with novelty. New labels can be generated as new affinity clusters appear, and posts with unusual combinations of features can be flagged for manual examination. Such unusual posts may indicate emerging events or other noteworthy changes in the social media environment. However, because algorithms generate the classification scheme, it can sometimes be difficult to interpret the classifications produced.

Social Verification

This technique examines the way other users interact with a post or an author to infer if a post is credible. This technique typically blends some combination of user- and post-based verification. User-based verification examines the other users with whom a user associates and scores the author as dubious if their associates are dubious. Signs of dubious associates often focus on indicators of bot accounts, such as inhuman posting volume, post timing that is inconsistent with stated location, bulk account creation, mass repeating, and abnormal emotive patterns. The exact best indicators vary over time as tactics evolve in the arms race between bot-creating trolls and the platform administrators that fight them. User-based verification works best when the anti-troll detection strategies are ahead of the troll tactics in use.

Post-based verification examines how other users react to a post. Signs of dubious posts include replies expressing incredulity or disagreement. They can also include signs that a post is generating an abnormal number of strong reactions but few affirming behaviors (such as reposting). This technique works best when the typical range of reactions to the posts are not too emotionally nuanced, because algorithmic measurement of emotion remains a difficult problem.

Blended Computer-Human Systems

This technique deploys algorithmic techniques to narrow down a large number of posts and then applies human intelligence to assessing the remainder. This technique strives to blend the relative strengths of machine and human intelligence. Machine intelligence excels at understanding statistical patterns and applying simple analysis rules quickly to a large body of data. Human intelligence excels at understanding behavioral patterns and conducting deep analysis of a small body of data.[30] This technique is viable when algorithmic methods can narrow down the data pool to a subset that available analyst resources can then process. It is cost-effective when the analysis of the remaining data pool is easy for human intelligence but difficult for machine intelligence.

---

[30] More formally, human intelligence excels at put-yourself-in-their-shoes empathetic behavioral modeling, including both "in that situation, I'd feel . . . " complex emotional processing and recursive "I think that you think that I think" reasoning.

References

ABC News, "Russia and Syria Blame Rebels for Suspected Poison Gas Attack in Aleppo," November 25, 2018. As of December 13, 2020: https://www.abc.net.au/news/2018-11-26/russia-bombs-syria-rebels-after-suspected-aleppo-gas-attack/10553338

Ahmad, Kashif, Konstantin Pogorelov, Michael Riegler, Nicola Conci, and Pål Halvorsen, "Social Media and Satellites," *Multimedia Tools and Applications*, Vol. 78, No. 3, February 2019, pp. 2837–2875.

Alexa, "The Top 500 Sites on the Web," webpage, undated. As of June 25, 2020: https://www.alexa.com/topsites/countries

Alkhodair, Sarah A., Steven H. H. Ding, Benjamin C. M. Fung, and Junqiang Liu, "Detecting Breaking News Rumors of Emerging Topics in Social Media," *Information Processing and Management*, Vol. 57, No. 2, March 2020.

Almukhtar, Sarah, "Most Chemical Attacks in Syria Get Little Attention. Here Are 34 Confirmed Cases," *New York Times*, April 13, 2018. As of November 12, 2020: https://www.nytimes.com/interactive/2018/04/13/world/middleeast/yria-chemical-attacks-maps-history.html

American Psychiatric Association, *Diagnostic and Statistical Manual of Mental Disorders*, 5th ed., Washington, D.C., 2013.

Associated Press, "Timeline of Chemical Weapons Attacks in Syria," April 10, 2018. As of November 11, 2020: https://apnews.com/article/74085b6b92c446678cfe704ee352c5ba

Babić Rosario, Ana, Francesca Sotgiu, Kristine De Valck, and Tammo Bijmolt, "The Effect of Electronic Word of Mouth on Sales: A Meta-Analytic Review of Platform, Product, and Metric Factors," *Journal of Marketing Research*, Vol. 53, No. 3, June 2016, pp. 297–318.

Badawy, Adam, Aseel Addawood, Kristina Lerman, and Emilio Ferrara, "Characterizing the 2016 Russian IRA Influence Campaign," *Social Network Analysis and Mining*, Vol. 9, No. 1, 2019, article 31.

Barnard, Anne, and Hwaida Saad, "Despite U.N. Cease-Fire, Syrian Forces Begin New Attacks in Rebel-Held Enclave," *New York Times*, February 25, 2018. As of November 12, 2020:
https://www.nytimes.com/2018/02/25/world/middleeast/syria-united-nations-ceasefire-ghouta.html

BBC News, "Syria War: 'Chlorine Attack' on Rebel-Held Idlib Town," webpage, February 5, 2018. As of November 12, 2020:
https://www.bbc.com/news/world-middle-east-42944033

Bello-Orgaz, Gema, Jason J. Jung, and David Camacho, "Social Big Data: Recent Achievements and New Challenges," *Information Fusion*, Vol. 28, March 2016, pp. 45–59.

Benigni, Matthew, Kenneth Joseph, and Kathleen M. Carley, "Bot-ivism: Assessing Information Manipulation in Social Media Using Network Analytics," in Nitin Agarwal, Nima Dokoohaki, and Serpil Tokdemir, eds., *Emerging Research Challenges and Opportunities in Social Network Analysis and Mining*, Cham, Switzerland: Springer, 2019, pp. 19–42.

Beskow, David, and Kathleen M. Carley, "Bot Conversations Are Different: Leveraging Network Metrics for Bot Detection in Twitter," in *2018 IEEE/ACM International Conference on Advances in Social Networks Analysis and Mining (ASONAM) Proceedings*, Barcelona, Spain, August 28–31, 2018, pp. 825–832.

Bodine-Baron, Elizabeth, Todd C. Helmus, Andrew Radin, and Elina Treyger, *Countering Russian Social Media Influence*, Santa Monica, Calif.: RAND Corporation, RR-2740-RC, 2018. As of June 25, 2020:
https://www.rand.org/pubs/research_reports/RR2740.html

Bradshaw, Samantha, and Philip N. Howard, *The Global Disinformation Order: 2019 Global Inventory of Organised Social Media Manipulation*, Oxford: Computational Propaganda Project at the Oxford Internet Institute, 2019. As of November 12, 2020:
https://comprop.oii.ox.ac.uk/wp-content/uploads/sites/93/2019/09/CyberTroop-Report19.pdf

Bride, Brian E., Margaret M. Robinson, Bonnie Yegidis, and Charles R. Figley, "Development and Validation of the Secondary Traumatic Stress Scale," *Research on Social Work Practice*, Vol. 14, No. 1, January 2004, pp. 27–35.

Central Intelligence Agency, "The World Factbook," webpage, undated. As of June 25, 2020:
https://www.cia.gov/library/publications/the-world-factbook

Chaabane, Abdelberi, Terence Chen, Mathieu Cunche, Emiliano De Cristofaro, Arik Friedman, and Mohamed Ali Kaafar, "Censorship in the Wild: Analyzing Internet Filtering in Syria," *IMC '14: Proceedings of the 2014 Conference on Internet Measurement Conference*, New York: Association for Computing Machinery, 2014, pp. 285–298.

Choi, Hyeok-Jun, and Cheong Hee Park, "Emerging Topic Detection in Twitter Stream Based on High Utility Pattern Mining," *Expert Systems with Applications*, Vol. 115, January 2019, pp. 27–36.

Cleverdon, Cyril W., "On the Inverse Relationship of Recall and Precision," *Journal of Documentation*, Vol. 28, No. 3, 1972, pp. 195–201.

Collins, Rebecca L., Eunice C. Wong, Joshua Breslau, M. Audrey Burnam, Matthew Cefalu, and Elizabeth Roth, "Social Marketing of Mental Health Treatment: California's Mental Illness Stigma Reduction Campaign," *American Journal of Public Health*, Vol. 109, Supplement 3, June 2019, pp. S228–S235.

Corder, Mike, "Watchdog: Chlorine Used in Syrian Town of Saraqeb," Associated Press, May 16, 2018.

Craig, Carlton D., and Ginny Sprang, "Compassion Satisfaction, Compassion Fatigue, and Burnout in a National Sample of Trauma Treatment Therapists," *Anxiety Stress Coping*, Vol. 23, No. 3, May 2010, pp. 319–339.

DataReportal, "Digital 2019: Syria," DataReportal webpage, January 31, 2019. As of November 12, 2020:
https://datareportal.com/reports/digital-2019-syria

Del Vicario, Michela, Alessandro Bessi, Fabiana Zollo, Fabio Petroni, Antonio Scala, Guido Caldarelli, H. Eugene Stanley, and Walter Quattrociocchi, "The Spreading of Misinformation Online," *Proceedings of the National Academy of Sciences of the United States of America*, Vol. 113, No. 3, January 2016, pp. 554–559.

Demchenko, Yuri, Canh Ngo, and Peter Membrey, "Architecture Framework and Components for the Big Data Ecosystem," Draft Version 0.2, September 12, 2013. As of November 12, 2020:
http://citeseerx.ist.psu.edu/viewdoc/download;jsessionid=88652198AADB90A0D CD1BBB61EF394DF?doi=10.1.1.670.8078&rep=rep1&type=pdf

Eberhard, David M., Gary F. Simons, and Charles D. Fennig, eds., "Ethnologue: Languages of the World," 23rd ed., website, Dallas, Tex.: SIL International, undated. As of June 25, 2020:
https://www.ethnologue.com

Edwards, Donna, Paul Krauter, David Franco, and Mark Tucker, *Key Planning Factors: For Recovery from a Chemical Warfare Agent Incident, Summer 2019*, Washington, D.C.: U.S. Department of Homeland Security, Science and Technology, Summer 2012. As of June 25, 2020:
https://www.fema.gov/media-library-data/20130726-1910-25045-7886/10_rrkp _key_planning_factors_chemical_incident.pdf

Entous, Adam, Ellen Nakashima, and Greg Jaffe, "Kremlin Trolls Burned Across the Internet as Washington Debated Options," *Washington Post*, December 25, 2017. As of November 18, 2020:
https://www.washingtonpost.com/world/national-security/kremlin-trolls-burned-across-the-internet-as-washington-debated-options/2017/12/23/e7b9dc92-e403-11e7-ab50-621fe0588340_story.html

Fan, Chao, and Ali Mostafavi, "A Graph-Based Method for Social Sensing of Infrastructure Disruptions in Disasters," *Computer-Aided Civil and Infrastructure Engineering*, Vol. 34, No. 12, May 2019, pp. 1055–1070.

Ferrara, Emilio, Onur Varol, Clayton Davis, Filippo Menczer, and Alessandro Flammini, "The Rise of Social Bots," *Communications of the ACM*, Vol. 59, No. 7, July 2016, pp. 96–104.

Figley, Charles R., *Compassion Fatigue: Coping with Secondary Traumatic Stress Disorder in Those Who Treat the Traumatized*, New York: Brunner/Mazel, 1995.

Flickr, website, undated. As of December 15, 2020:
https://www.flickr.com/

Fortin, Jacey, "When Disaster Hits and Landlines Fail, Social Media Is a Lifeline," *New York Times*, 2017. As of June 24, 2020:
https://www.nytimes.com/2017/09/23/us/whatsapp-zello-hurricane-earthquakes.html

Freedom House, "Internet Freedom Status," webpage, undated. As of June 25, 2020:
https://freedomhouse.org/explore-the-map?type=fotn&year=2019

Gendronneau, Cloé, Arkadiusz Wisniowski, Dilek Yildiz, Emilio Zagheni, Lee Florio, Yuan Hsiao, Martin Stepanek, Ingmar Weber, Guy Abel, and Stijn Hoorens, *Measuring Labour Mobility and Migration Using Big Data: Exploring the Potential of Social-Media Data for Measuring EU Mobility Flows and Stocks of EU Movers*, Brussels: European Commission, 2019. As of November 18, 2020:
https://www.rand.org/pubs/external_publications/EP68038.html

Google, "Google Maps," undated. As of December 15, 2020:
https://www.google.com/maps

Gordon, Michael, and Manfred Kochen, "Recall-Precision Trade-Off: A Derivation," *Journal of the American Society for Information Science*, Vol. 40, No. 3, May 1989, pp. 145–151.

Hasan, Mahmud, Mehmet A. Orgun, and Rolf Schwitter, "Real-Time Event Detection from the Twitter Data Stream Using the TwitterNews+ Framework," *Information Processing and Management*, Vol. 56, No. 3, May 2019, pp. 1146–1165.

Headquarters, Department of the Army, *Multi-Service Doctrine for Chemical, Biological, Radiological, and Nuclear Operations*, Washington, D.C., July 2011. As of June 25, 2020:
https://apps.dtic.mil/dtic/tr/fulltext/u2/a550361.pdf

Helmus, Todd C., Elizabeth Bodine-Baron, Andrew Radin, Madeline Magnuson, Joshua Mendelsohn, William Marcellino, Andriy Bega, and Zev Winkelman, *Russian Social Media Influence: Understanding Russian Propaganda in Eastern Europe*, Santa Monica, Calif.: RAND Corporation, RR-2237-OSD, 2018. As of June 25, 2020:
https://www.rand.org/pubs/research_reports/RR2237.html

Hensel, Jennifer M., Carlos Ruiz, Caitlin Finney, and Carolyn S. Dewa, "Meta-Analysis of Risk Factors for Secondary Traumatic Stress in Therapeutic Work with Trauma Victims," *Journal of Traumatic Stress*, Vol. 28, No. 2, April 2015, pp. 83–91.

Higgins, Eliot, "Images from the January 22nd 2018 Chlorine Attack in Douma, Damascus," *Bellingcat* blog, February 5, 2018. As of November 12, 2020:
https://www.bellingcat.com/news/mena/2018/02/05/images-january-22nd-2018-chlorine-attack-douma-damascus/

House Armed Services Committee, *Statement of Vayl Oxford*, testimony before the Subcommittee on Intelligence and Emerging Threats and Capabilities, Washington, D.C., February 11, 2020. As of December 13, 2020:
https://www.congress.gov/116/meeting/house/110440/witnesses/HHRG-116-AS26-Bio-OxfordV-20200211.pdf

Hubbard, Ben, "Syria Urges U.N. to Condemn Rebels After Apparent Chemical Attack," *New York Times*, November 25, 2018. As of November 12, 2020:
https://www.nytimes.com/2018/11/25/world/middleeast/syria-chemical-weapons-un.html

Humud, Carla E., Christopher M. Blanchard, and Mary Beth D. Nikitin, *Armed Conflict in Syria: Overview and U.S. Response*, Washington, D.C.: Congressional Research Service, RL33487, January 6, 2017. As of November 12, 2020:
https://apps.dtic.mil/dtic/tr/fulltext/u2/1024469.pdf

Jacobsen, Kathryn H., A. Alonso Aguirre, Charles L. Bailey, Ancha V. Baranova, Andrew T. Crooks, Arie Croitoru, Paul L. Delamater, Jhumka Gupta, Kylene Kehn-Hall, Aarthi Narayanan, Mariaelena Pierobon, Katherine E. Rowan, J. Reid Schwebach, Padmanabhan Seshaiyer, Dann M. Sklarew, Anthony Stefanidis, and Peggy Agouris, "Lessons from the Ebola Outbreak: Action Items for Emerging Infectious Disease Preparedness and Response," *Ecohealth*, Vol. 13, No. 1, March 2016, pp. 200–212.

Jingchao Yang, Manzhu Yu, Han Qin, Mingyu Lu, and Chaowei Yang, "A Twitter Data Credibility Framework—Hurricane Harvey as a Use Case," *ISPRS International Journal of Geo-Information*, Vol. 8, No. 3, 2019, p. 111.

Joint Publication 3-41, *Chemical, Biological, Radiological, and Nuclear Response*, Washington, D.C.: Joint Chiefs of Staff, September 9, 2016. As of June 25, 2020: https://www.jcs.mil/Portals/36/Documents/Doctrine/pubs/jp3_41.pdf

Kase, Sue E., Elizabeth K. Bowman, Tanvir Al Amin, and Tarek Abdelzaher, "Exploiting Social Media for Army Operations: Syrian Civil War Use Case," *Proceedings of SPIE (the International Society for Optical Engineering)*, Vol. 9122, July 2014.

Kragh, Martin, and Sebastian Åsberg, "Russia's Strategy for Influence Through Public Diplomacy and Active Measures: The Swedish Case," *Journal of Strategic Studies*, Vol. 40, No. 6, 2017, pp. 773–816.

Kumar, Ashish, Ram Bezawada, Rishika Rishika, Ramkumar Janakiraman, and P. K. Kannan, "From Social to Sale: The Effects of Firm-Generated Content in Social Media on Customer Behavior," *Journal of Marketing*, Vol. 80, No. 1, January 2016, pp. 7–25.

Lamberton, Cait, and Andrew T. Stephen, "A Thematic Exploration of Digital, Social Media, and Mobile Marketing: Research Evolution from 2000 to 2015 and an Agenda for Future Inquiry," *Journal of Marketing*, Vol. 80, No. 6, November 2016, pp. 146–172.

Lessenberry, Brian, "Intelligence Integration and the Syrian CW Threat," Center for Strategic and International Studies webpage, February 18, 2015. As of February 12, 2020:
https://www.csis.org/analysis/intelligence-integration-and-syrian-cw-threat

Lister, Charles R., *The Syrian Jihad: Al-Qaeda, the Islamic State and the Evolution of an Insurgency*, New York: Oxford University Press, 2016.

Marcellino, William, Krystyna Marcinek, Stephanie Pezard, and Miriam Matthews, *Detecting Malign or Subversive Information Efforts over Social Media: Scalable Analytics for Early Warning*, Santa Monica, Calif.: RAND Corporation, RR-4192-EUCOM, 2020. As of June 22, 2020:
https://www.rand.org/pubs/research_reports/RR4192.html

Marcellino, William, Meagan L. Smith, Christopher Paul, and Lauren Skrabala, *Monitoring Social Media: Lessons for Future Department of Defense Social Media Analysis in Support of Information Operations*, Santa Monica, Calif.: RAND Corporation, RR-1742-OSD, 2017. As of June 25, 2020:
https://www.rand.org/pubs/research_reports/RR1742.html

Maslach Christina, Susan E. Jackson, and Michael P. Leiter, *Maslach Burnout Inventory*, 3rd ed., Palo Alto, Calif.: Consulting Psychologists Press, 1986.

Maslach Christina, and Michael P. Leiter, "New Insights into Burnout and Health Care: Strategies for Improving Civility and Alleviating Burnout," *Medical Teacher*, Vol. 39, No. 2, 2017, pp. 160–163.

Mattis, Jim, *Summary of the 2018 National Defense Strategy of the United States of America*, Washington, D.C.: U.S. Department of Defense, 2018. As of November 12, 2020:
https://dod.defense.gov/Portals/1/Documents/pubs/2018-National-Defense-Strategy-Summary.pdf

McCulloh, Ian, and Kathleen M. Carley, "Detecting Change in Longitudinal Social Networks," *Journal of Social Structure*, Vol. 12, No. 1, 2011, pp. 1–37.

MediaWell, various analyses of disinformation, webpage. As of June 25, 2020:
https://mediawell.ssrc.org

Mendoza, Marcelo, Bárbara Poblete, and Ignacio Valderrama, "Nowcasting Earthquake Damages with Twitter," *EPJ Data Science*, Vol. 8, No. 3, 2019, pp. 1–23.

Mønsted, Bjarke, Piotr Sapieżyński, Emilio Ferrara, and Sune Lehmann, "Evidence of Complex Contagion of Information in Social Media: An Experiment Using Twitter Bots," *PloS One*, Vol. 12, No. 9, September 2017, pp. 1–12.

Nair, Ajay, "Chemical Attacks in Syria: A Deadly History," *Sky News,* April 14, 2018. As of November 12, 2020:
https://news.sky.com/story/chemical-attacks-in-syria-a-deadly-history-11323747

Naslund, John A., Kelly A. Aschbrenner, Lisa A. Marsch, and Stephen J. Bartels, "The Future of Mental Health Care: Peer-to-Peer Support and Social Media," *Epidemiology and Psychiatric Sciences*, Vol. 25, No. 2, April 2016, pp. 113–122.

National Child Traumatic Stress Network, "Secondary Traumatic Stress," webpage, undated. As of November 18, 2020:
https://www.nctsn.org/trauma-informed-care/secondary-traumatic-stress

National Child Traumatic Stress Network, "Taking Care of Yourself," fact sheet, 2018. As of November 18, 2020:
https://www.nctsn.org/print/2038

National Child Traumatic Stress Network, "Secondary Traumatic Stress Core Competencies in Trauma-Informed Supervision Self-Rating Tool," special resource, 2019. As of November 18, 2020:
https://www.nctsn.org/print/2070

Nazir, Faria, Mustansar Ali Ghazanfar, Muazzam Maqsood, Farhan Aadil, Seungmin Rho, and Irfan Mehmood, "Social Media Signal Detection Using Tweets Volume, Hashtag, and Sentiment Analysis," *Multimedia Tools and Applications*, Vol. 78, No. 3, February 2019, pp. 3553–3586.

Newell, Jason M., Debra Nelson-Gardell, and Gordon MacNeil, "Clinician Responses to Client Traumas: A Chronological Review of Constructs and Terminology," *Trauma, Violence, and Abuse*, Vol. 17, No. 3, July 2016, pp. 306–313.

Ni, Ming, Qing He, and Jing Gao, "Forecasting the Subway Passenger Flow Under Event Occurrences with Social Media," *IEEE Transactions on Intelligent Transportation Systems*, Vol. 18, No. 6, June 2017, pp. 1623–1632.

Nolasco, Diogo, and Jonice Oliveira, "Subevents Detection Through Topic Modeling in Social Media Posts," *Future Generation Computer Systems*, Vol. 93, April 2019, pp. 290–303.

Obama, Barack, "Remarks by the President to the White House Press Corps," webpage, August 20, 2012. As of November 12, 2020:
https://obamawhitehouse.archives.gov/the-press-office/2012/08/20/remarks-president-white-house-press-corps

Organization for the Prohibition of Chemical Weapons, "Syria and the OPCW," webpage, 2020. As of December 13, 2020:
https://www.opcw.org/media-centre/featured-topics/syria-and-opcw

Oxford, Vayl S., "Countering Threat Networks to Deter, Compete, and Competition Below Armed Conflict with Revisionist Powers," *Joint Forces Quarterly*, No. 95, 4th Quarter 2019, pp. 79–80.

Pai, Anushka, Alina M. Suris, and Carol S. North, "Posttraumatic Stress Disorder in the *DSM-5*: Controversy, Change, and Conceptual Considerations," *Behavioral Sciences*, Vol. 7, No. 1, March 2017, p. 7.

Pierce, Carrie E., Khaled Bouri, Carol Pamer, Scott Proestel, Harold W. Rodriguez, Hoa Van Le, Clark C. Freifeld, John S. Brownstein, Mark Walderhaug, I. Ralph Edwards, and Nabarun Dasgupta, "Evaluation of Facebook and Twitter Monitoring to Detect Safety Signals for Medical Products: An Analysis of Recent FDA Safety Alerts," *Drug Safety*, Vol. 40, No. 4, 2017, pp. 317–331.

Qian, Xueming, Mingdi Li, Yayun Ren, and Shuhui Jiang, "Social Media Based Event Summarization by User–Text–Image Co-Clustering," *Knowledge-Based Systems*, Vol. 164, No. 15, January 2019, pp. 107–121.

Reporters Without Borders, "World Press Freedom Index: Ranking 2020," webpage, 2020. As of June 25, 2020:
https://rsf.org/en/ranking

Reuters, "Rescuers in Rebel-Held Syrian Area Accuse Government of Gas Attack," January 22, 2018. As of November 12, 2020:
https://www.reuters.com/article/us-mideast-crisis-syria-ghouta/rescuers-in-rebel-held-syrian-area-accuse-government-of-gas-attack-idUSKBN1FB135

Reuters, "Turkish Army Hit Village in Syria's Afrin with Suspected Gas: Kurdish YPG, Observatory," February 16, 2018. As of November 12, 2020:
https://www.reuters.com/article/us-mideast-crisis-syria-turkey-afrin/turkish-army-hit-village-in-syrias-afrin-with-suspected-gas-kurdish-ypg-observatory-idUSKCN1G02JE

Rosser, Julian F., Didier G. Leibovici, and Margaret J. Jackson, "Rapid Flood Inundation Mapping Using Social Media, Remote Sensing and Topographic Data," *Natural Hazards*, Vol. 87, No. 1, 2017, pp. 103–120.

Saeed, Zafar, Rabeeh Ayaz Abbasi, Onaiza Maqbool, Abida Sadaf, Imran Razzak, Ali Daud, Naif Radi Aljohani, and Guandong Xu, "What's Happening Around the World? A Survey and Framework on Event Detection Techniques on Twitter," *Journal of Grid Computing*, Vol. 17, No. 2, June 2019, pp. 279–312.

Sanders-Zakre, Alicia, "Timeline of Syrian Chemical Weapons Activity, 2012–2019," Arms Control Association website, March 2019. As of November 11, 2020: https://perma.cc/LGA6-BKC2

Schehl, Matthew, "NPS Team Turns to Machine Learning to Predict Social Unrest," Naval Postgraduate School webpage, January 16, 2019. As of August 20, 2020:
https://nps.edu/-/nps-team-turns-to-machine-learning-to-predict-social-unrest

Schneider, Tobias, and Theresa Lütkefend, *Nowhere to Hide: The Logic of Chemical Weapons Use in Syria*, Berlin: Global Public Policy Institute, February 2019.

Schwab, Katharine, "The Internet Isn't Available in Most Languages," *The Atlantic*, November 2015. As of June 24, 2020:
https://www.theatlantic.com/technology/archive/2015/11/the-internet-isnt-available-in-most-languages/417393

Şerban, Ovidiu, Nicholas Thapen, Brendan Maginnis, Chris Hankin, and Virginia Foot, "Real-Time Processing of Social Media with SENTINEL: A Syndromic Surveillance System Incorporating Deep Learning for Health Classification," *Information Processing and Management*, Vol. 56, No. 3, May 2019, pp. 1166–1184.

Singh, Jyoti Prakash, Yogesh K. Dwivedi, Nripendra P. Rana, Abhinav Kumar, and Kawaljeet Kaur Kapoor, "Event Classification and Location Prediction from Tweets During Disasters," *Annals of Operations Research*, Vol. 283, No. 1, 2019, pp. 737–757.

Sirgany, Sarah, and Eyad Kourdi, "Dozens Injured in Toxic Gas Attack on Aleppo, Syria, Reports Say," CNN, November 26, 2018. As of November 12, 2020: https://www.cnn.com/2018/11/25/middleeast/syria-gas-attacks/index.html

Stamm, B. Hudnall, *Secondary Traumatic Stress: Self-Care Issues for Clinicians, Researchers, and Educators*, 2nd ed., Lutherville, Md.: Sidran Press, 1999.

Stamm, B. Hudnall, *Professional Quality of Life: Compassion Satisfaction and Fatigue Version 5 (ProQOL)*, 2009. As of November 12, 2020:
https://socialwork.buffalo.edu/content/dam/socialwork/home/self-care-kit/compassion-satisfaction-and-fatigue-stamm-2009.pdf

Ting, Laura, Jodi Jacobson, and Sara Sanders, "Current Levels of Perceived Stress Among Mental Health Social Workers Who Work with Suicidal Clients," *Social Work*, Vol. 56, No. 4, 2011, pp. 327–336.

Twenge, Jean M., Thomas E. Joiner, Megan L. Rogers, and Gabrielle N. Martin, "Increases in Depressive Symptoms, Suicide-Related Outcomes, and Suicide Rates Among U.S. Adolescents After 2010 and Links to Increased New Media Screen Time," *Clinical Psychological Science*, Vol. 6, No. 1, 2018, pp. 3–17.

Twitter, "Developer: Rate Limits, Overview, Standard API Rate Limits Per Window," webpage, undated. As of June 25, 2020:
https://developer.twitter.com/en/docs/basics/rate-limits

Twitter, "Developer: Developer Agreement and Policy," webpage, March 10, 2020. As of June 25, 2020:
https://developer.twitter.com/en/developer-terms/agreement-and-policy

United Nations Human Rights Council, "Chemical Weapons Attacks Documented by the Independent International Commission of Inquiry on the Syrian Arab Republic," January 15, 2018. As of November 12, 2020:
https://www.ohchr.org/SiteCollectionImages/Bodies/HRCouncil/IICISyria/COISyria_ChemicalWeapons.jpg

Uras, Umut, "Turkey Official Denies Use of Chemical Weapons in Afrin," *Al Jazeera*, February 18, 2018. As of November 11, 2020:
https://www.aljazeera.com/news/2018/2/18/turkey-official-denies-use-of-chemical-weapons-in-afrin

U.S. Department of Defense, *The Militarily Critical Technologies List Part II: Weapons of Mass Destruction Technologies*, Washington, D.C.: Office of the Under Secretary of Defense for Acquisition and Technology, February 1998. As of November 12, 2020:
https://fas.org/irp/threat/mctl98-2/mctl98-2.pdf

U.S. Department of Homeland Security, "Chemical Attack Fact Sheet: Warfare Agents, Industrial Chemicals, and Toxins," Washington, D.C.: National Academy of Sciences, 2004. As of November 12, 2020:
https://www.dhs.gov/sites/default/files/publications/prep_chemical_fact_sheet.pdf

Van Stekelenborg, John, Johan Ellenius, Simon Maskell, Tomas Bergvall, Ola Caster, Nabarun Dasgupta, Juergen Dietrich, Sara Gama, Davis Lewis, Victoria Newbould, Sabine Brosch, Carrie E. Pierce, Gregory Powell, Alicia Ptaszyńska-Neophytou, Antoni F. Z. Wiśniewski, Phil Tregunno, G. Niklas Norén, and Munir Pirmohamed, "Recommendations for the Use of Social Media in Pharmacovigilance: Lessons from IMI WEB-RADR," *Drug Safety*, Vol. 42, No. 12, December 2019, pp. 1393–1407.

Vishwakarma, Dinesh Kumar, Deepika Varshney, and Ashima Yadav, "Detection and Veracity Analysis of Fake News via Scrapping and Authenticating the Web Search," *Cognitive Systems Research*, Vol. 58, December 2019, pp. 217–229.

Volkova, Svitlana, Kyle Shaffer, Jin Yea Jang, and Nathan Hodas, "Separating Facts from Fiction: Linguistic Models to Classify Suspicious and Trusted News Posts on Twitter," in Regina Barzilay and Min-Yen Kan, eds., *Proceedings of the 55th Annual Meeting of the Association for Computational Linguistics (*Vol. 2: *Short Papers)*, July 2017, pp. 647–653.

Vomfell, Lara, Wolfgang Karl Härdle, and Stefan Lessmann, "Improving Crime Count Forecasts Using Twitter and Taxi Data," *Decision Support Systems*, Vol. 113, September 2018, pp. 73–85.

Wang, Zhen, Chris T. Bauch, Samit Bhattacharyya, Alberto d'Onofrio, Piero Manfredi, Matjaz Perc, Nicola Perra, Marcel Salathé, and Dawei Zhao, "Statistical Physics of Vaccination," *Physics Reports*, Vol. 664, November 2016, pp. 1–113.

Warren, Rossalyn, "The Hidden Victims of Repression—How Activists and Reporters Can Protect Themselves from Secondary Trauma," Amnesty International website, February 20, 2019. As of November 18, 2020: https://www.amnesty.org/en/latest/news/2019/02/how-activists-and-reporters-can-protect-themselves-from-secondary-trauma

Warren, T. Camber, "Not by the Sword Alone: Soft Power, Mass Media, and the Production of State Sovereignty," *International Organization*, Vol. 68, No. 1, January 2014, pp. 111–141.

We Are Social, "Digital in 2020," webpage, undated. As of June 25, 2020: https://wearesocial.com/digital-2020

Wenger, Jennie W., Heather Krull, Elizabeth Bodine-Baron, Eric V. Larson, Joshua Mendelsohn, Tepring Piquado, and Christine Anne Vaughan, *Social Media and the Army: Implications for Outreach and Recruiting*, Santa Monica, Calif.: RAND Corporation, RR-2686-A, 2019. As of June 25, 2020: https://www.rand.org/pubs/research_reports/RR2686.html

Whittaker, Zack, "Surveillance and Censorship: Inside Syria's Internet," *CBS News*, December 12, 2013. As of June 24, 2020: https://www.cbsnews.com/news/surveillance-and-censorship-inside-syrias-internet

Wigert, Ben, and Sangeeta Agrawal, "Employee Burnout, Part 1: The 5 Main Causes," Gallup Workplace webpage, July 12, 2018. As of June 24, 2020: https://www.gallup.com/workplace/237059/employee-burnout-part-main-causes.aspx

Williams, Matthew L., and Pete Burnap, "Cyberhate on Social Media in the Aftermath of Woolwich: A Case Study in Computational Criminology and Big Data," *British Journal of Criminology*, Vol. 56, No. 2, March 2016, pp. 211–238.

Williams, Matthew L., Pete Burnap, Amir Javed, Han Liu, and Sefa Ozalp, "Hate in the Machine: Anti-Black and Anti-Muslim Social Media Posts as Predictors of Offline Racially and Religiously Aggravated Crime," *British Journal of Criminology*, Vol. 60, No. 1, January 2020, pp. 93–117.

Winter, Stephan, Miriam J. Metzger, and Andrew J. Flanagin, "Selective Use of News Cues: A Multiple-Motive Perspective on Information Selection in Social Media Environments," *Journal of Communication*, Vol. 66, No. 4, August 2016, pp. 669–693.

Woods, Heather Cleland, and Holly Scott, "#Sleepyteens: Social Media Use in Adolescence Is Associated with Poor Sleep Quality, Anxiety, Depression and Low Self-Esteem," *Journal of Adolescence*, Vol. 51, August 2016, pp. 41–49.

World Bank, "Access to Electricity (% of Population)," webpage, undated. As of June 25, 2020:
https://data.worldbank.org/indicator/EG.ELC.ACCS.ZS

World Bank, "Individuals Using the Internet (% of Population)," webpage, undated. As of June 25, 2020:
https://data.worldbank.org/indicator/IT.NET.USER.ZS

World Bank, "Mobile Cellular Subscriptions (per 100 People)," webpage, undated. As of June 25, 2020:
https://data.worldbank.org/indicator/IT.CEL.SETS.P2

World Health Organization, "Public Health Preparedness and Response," in *Public Health Response to Biological and Chemical Weapons: WHO Guidance*, 2nd ed., Geneva, 2004. As of June 25, 2020:
https://www.who.int/csr/delibepidemics/chapter4.pdf

Xiang, Zheng, Qianzhou Du, Yufeng Ma, and Weiguo Fan, "A Comparative Analysis of Major Online Review Platforms: Implications for Social Media Analytics in Hospitality and Tourism," *Tourism Management*, Vol. 58, February 2017, pp. 51–65.

Xu, Zheng, Yunhuai Liu, Junyu Xuan, Haiyan Chen, and Lin Mei, "Crowdsourcing Based Social Media Data Analysis of Urban Emergency Events," *Multimedia Tools and Applications*, Vol. 76, No. 9, 2017, pp. 11567–11584.

Yang, Dingqi, Terence Heaney, Alberto Tonon, Leye Wang, and Philippe Cudré-Mauroux, "CrimeTelescope: Crime Hotspot Prediction Based on Urban and Social Media Data Fusion," *World Wide Web*, Vol. 21, 2018, pp. 1323–1347.

Yeung, Douglas, Sarah A. Nowak, Sohaela Amiri, Aaron C. Davenport, Emily Hoch, Kelly Klima, and Colleen M. McCullough, *U.S. Coast Guard Emergency Response and Disaster Operations: Using Social Media for Situational Awareness*, Santa Monica, Calif.: RAND Corporation, RR-4296-DHS, 2020. As of November 12, 2020:
https://www.rand.org/pubs/research_reports/RR4296.html

YouTube, "YouTube API Services Terms of Service: Usage and Quotas," webpage, last updated August 27, 2020. As of November 12, 2020:
https://developers.google.com/youtube/v3/getting-started#quota

Zhang, Chao, Guangyu Zhou, Quan Yuan, Honglei Zhuang, Yu Zheng, Lance Kaplan, Shaowen Wang, and Jiawei Han, "Geoburst: Real-Time Local Event Detection in Geo-Tagged Tweet Streams," in *SIGIR '16: Proceedings of the 39th International ACM SIGIR Conference on Research and Development in Information Retrieval*, July 2016, pp. 513–522.

Zhang, Xichen, and Ali A. Ghorbani, "An Overview of Online Fake News: Characterization, Detection, and Discussion," *Information Processing and Management*, Vol. 57, No. 2, March 2020.

Zheng Xu, Hui Zhang, Chuanping Hu, Lin Mei, Junyu Xuan, Kim-Kwang Raymond Choo, Vjayan Sugumaran, and Yiwei Zhu, "Building Knowledge Base of Urban Emergency Events Based on Crowdsourcing of Social Media," *Concurrency and Computation: Practice and Experience*, Vol. 28, No. 15, October 2016, pp. 4038–4052.

Zheng Xu, Lin Mei, Zhihan Lv, Chuaping Hu, Xiangfeng Luo, Hui Zhang, and Yunhuai Liu, "Multi-Modal Description of Public Safety Events Using Surveillance and Social Media," *IEEE Transactions on Big Data*, Vol. 5, No. 4, October–December 2019, pp. 529–539.

Zhenguo Yang, Qing Li, Wenyin Liu, and Jianming Lv, "Shared Multi-View Data Representation for Multi-Domain Event Detection," *IEEE Transactions on Pattern Analysis and Machine Intelligence*, Vol. 42, No. 5, January 2019, pp. 1243–1256.

Zhicheng, Shi, and Lilian S. C. Pun-Cheng, "Spatiotemporal Data Clustering: A Survey of Methods," *ISPRS International Journal of Geo-Information*, Vol. 8, No. 3, 2019, p. 112.

Zubiaga, Arkaitz, Ahmet Aker, Kalina Bontcheva, Maria Liakata, and Rob Procter, "Detection and Resolution of Rumours in Social Media: A Survey," *ACM Computing Surveys*, Vol. 51, No. 2, February 2018, pp. 1–36.

Zubiaga, Arkaitz, Maria Liakata, and Rob Procter, "Exploiting Context for Rumour Detection in Social Media," *Social Informatics*, 2017, pp. 109–123.

Zubiaga, Arkaitz, Maria Liakata, Rob Procter, Geraldine Wong Sak Hoi, and Peter Tolmie, "Analysing How People Orient To and Spread Rumours in Social Media by Looking at Conversational Threads," *PloS One*, Vol. 11, No. 3, 2016.